Upstream Kanban

Anna Radzikowska

BLUE HOLE PRESS Chicago, Illinois   Kanban University PRESS Bilbao, Spain

Blue Hole Press
Chicago, Illinois
www.blueholepress.com

Kanban University Press
Bilbao, Spain
https://kanban.university

Upstream Kanban

First edition print, April 2025

For any publishing enquiries please email contact@kanbanbooks.com for rights requests, customized editions, translation requests, licensing, bulk orders or other enquiries. Additional print copies of this and other Kanban publications can be purchased via https://kanbanbooks.com and regional distribution partners.

Access to this book will also be available online via kanban.plus where you will find additional learning resources.

For enquiries relating to Kanban University directly, please contact info@kanban.university.

ISBN   978-1-960442-18-5   color (a)
       978-1-960442-19-2   black and white(a)
       978-1-960442-20-8   color (l)
       978-1-960442-21-5   black and white (l)

Cover art copyright © Mauvius Group Inc, d.b.a. Blue Hole Press
Cover photo copyright © Mauvius Group Inc, d.b.a. Blue Hole Press
Cover design by Nastya Kondratova
Interior design by Vicki L. Rowland

Contents

Table of Figures

Preface

It was September 2023, and I stood on the Kanban Brazil stage answering questions after my conference talk. The time was almost up, and organizers allowed for the last question from the public.

"If you could do one thing differently, having the knowledge that you have, what would it be?" I wasn't prepared for that question, but my subconscious was prepared for the answer: "I would definitely start with learning more about the 'upstream.'" The audience laughed, as my talk was focused on improving the work and communication with clients using Upstream Kanban, and I had been teaching the *Kanban for Design and Innovation* class for more than three years already at the David J Anderson School of Management.

Preparation for this speech was a trip down memory lane, as I was "that annoying client" for development teams for most of my professional life. I spent thirteen years of my career working in the finance department at different companies, and there was always someone approaching me asking, "Can you tell me what exactly you would like us to build?" and, a few weeks later, "Can you test this stuff we prepared?" Although I was a business analyst and (user acceptance) tester, my identity was still strongly rooted in accounting and finance management. At some point, I played the role of classical project manager, officially hired as Senior RtR[1] Accountant.

Later (finally), I started working as a Business Analyst and Product Owner, still in the "requirements" zone of the processes. Why do I mention this? Because most of the time my colleagues and I kept struggling with the same problems over and over again: expanding scope; unrealistic delivery demands; constant requests for estimation, prioritization, and

---

[1] RtR, for Record to Report, is a finance and accounting management process that involves collecting, processing, and presenting accurate financial data. R2R provides strategic, financial, and operational feedback on the performance of the organization to inform management and other stakeholders (https://www.gartner.com/en/finance/glossary/record-to-report-r2r-).

reprioritization; defects blowing up in late phases of testing; frustrated team members; and disappointed clients.

Every time, we repeated to ourselves, "Next time we will do it better. We know what mistakes we made." Did we? You know the answer: No. The cycle started again and ended up the same way—until the last project, in which, having the support of great people working with me, I had a chance to do something in a DIFFERENT way, finally. The difference was the Kanban Method, and it eventually pushed me into the arms of Kanban for good.

In this book, however, I will not tell you the story of one project and ask you to believe that replicating what we did will save your process. Instead, I have collected multiple stories and examples of what people do in and with their upstream systems, which practices they use, and how those practices can bring positive change—and very often immediate benefits: successful communication with clients, building reliable expectations, and delivering what their clients need.

Anna Radzikowska
April 2025

Introduction

What Is Upstream Kanban and Why Do We Need It?

In the early days, the Kanban Method focused mostly on delivery teams: how to relieve people from being overburdened and improve the flow of work so that it was fast and predictable. Predictability came from low uncertainty: reducing rework, delay, and unnecessary tasks; eliminating work that gets discarded or aborted before delivery; improving delivery times; scheduling the delivery with greater predictability; and deferring commitment on work until there is sufficient certainty on what is needed and how to deliver it. During this time, the commitment point to the left of the delivery process was the horizon of Kanban implementations. Kanban systems spanned the process from commitment to delivery.

Around 2010, a couple of coaches and consultants working independently started experimenting with utilizing the concepts of the Kanban Method and adjusting them to the *terra incognita* between the *Pool of Work* and *Commitment*. This place was called Upstream or Discovery Kanban, and it was soon clear that there was immense value in exploring it further.

- What should we work on now?
- What can wait until later? And if later, when?
- What should we discard altogether?

Those questions, installed to prevent shallow work items from entering the delivery system, also became relevant for Upstream Kanban. Industries in which the goods are intangible struggle with the concept of limits. The upstream employees suffer from overburdening similarly to their downstream colleagues. It is easy to say yes to starting something. Committing makes customers happy: They have the impression that their work is in progress and that soon they will see the results. But "the bigger challenge is finishing something and doing so with predictability.

The more you say yes, the more you start, the less gets finished, the longer everything takes—and any predictability on delivery time dissolves away."[2]

One of the core concepts in Upstream Kanban is options. Having options means having a choice. Options help you manage the uncertainty embedded in knowledge work and ensure that you make the best choice available from all the possible . . . options. By introducing options—and deferring commitment—we make sure to always deliver what suits our customers best.

At each stage in the Upstream discovery—or ideation—process, there is the opportunity to discard a bad idea: It's too expensive; it's too hard technically; it will take too long; it has limited value or is generally undesirable to consumers. Upstream Kanban isn't so much about managing flow, as in Downstream Delivery Kanban systems. Instead, it is about marshaling options—having enough choices at the right time, without overburdening the system and the creative workers who generate those options by asking them to maintain too much work in progress.

Upstream Is a Funnel

Unlike the Downstream system of rather linear and sequential work, the better representation of the Upstream will be a funnel. We start with a high number of options, some of which will be discarded in the Discovery process.

The original diagram (Figure 1) shows the Upstream funnel and is a great metaphor for understanding the concept of narrowing down the number of options until the commitment point. The problem with the funnel metaphor is that it doesn't show discarding and shelving options; so, while it is easy to comprehend that we narrow down the options, an actual funnel does not throw anything away.

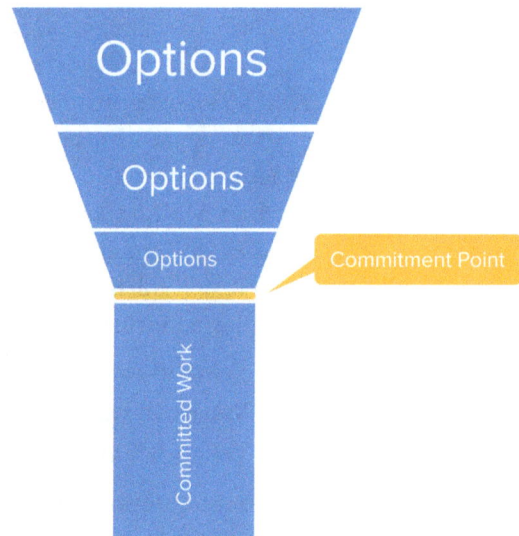

Figure 1 The options funnel as depicted by Patrick Steyaert in *Essential Upstream Kanban* (2018)

When explaining the concepts of the funnel and how to apply them to the system design, it has often been suggested to simply turn it 90 degrees to the left (kudos to

---

[2] David J Anderson, foreword to *Essential Upstream Kanban: Creating Value before Commitment* by Patrick Steyaert (Lean Kanban University Press, 2018), page vi.

Andreas Bartel, Accredited Kanban Trainer, coach, and consultant). This way, we depict a horizontal funnel that can be visualized on a more classically designed kanban board (Figure 2).

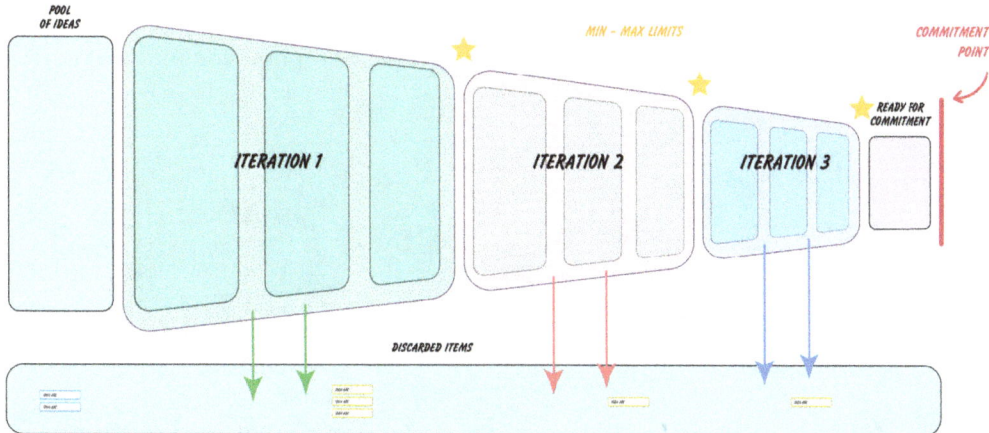

Figure 2 Horizontal funnel with discarded options

The discovery process involves analysis, synthesis, and decision points. The discarding process is forced by maximum WIP limits in each Upstream activity (which also supports relief from overburdening).

The characteristic element for Upstream Kanban is a minimum WIP limit. It is necessary always to have enough options from which to choose—sufficient choices to make the best commitments and facilitate the triage process.

By introducing minimum limits to signal replenishment, Upstream Kanban introduces a second kind of kanban signal card used in making physical, tangible goods. There are kanbans to limit inventory and prevent overburdening a system, and there are kanbans to signal replenishment. Read more about design considerations in Chapter 2, Designing Upstream Kanban.

In Upstream Kanban, we are discovering information, and doing so is still a service. The main difference between Upstream and Downstream is that Upstream allows for discarding ideas based on information discovered at each stage. The end product of Upstream is the final go or no-go decision that results in an eventual commitment.

What Does This Book Cover?

In this book, we examine:

- The concept of real options in Upstream Kanban and techniques for managing them

- Design elements of the Upstream Kanban system and different examples of designs
- Upstream feedback mechanisms
- Upstream metrics
- Managing backlogs based on the value of information
- Planning techniques using cost of delay, triage, and two-phase commitment

1

Introduction to Real Options

What Are Real Options and Why Are They Important to the Upstream Kanban Process?

The concept of real options comes from the financial management sector. The name *real option* refers to work involving tangible assets (such as machinery, land, buildings, and inventory) versus financial instruments.

Options exist when the decision maker has the *right*, but *not the obligation*, to *take a particular action*. These options add value by offering the flexibility to capitalize on opportunities as uncertainty gradually diminishes over time.

Real options are choices a company's management gives itself: the option to make, expand, change, or curtail projects based on changing economic, technological, or market conditions.[1]

The characteristics of real options include:

- The possibility, but not the obligation, to undertake certain business opportunities or investments by the company
- The possibility to expand, defer, redeploy, or abandon work entirely
- Their economic value, which financial analysts and corporate managers can use to inform their decisions

Real options embody flexibility in project development, serving as either a form of insurance or a means to quickly capitalize on favorable situations.

Real options represent opportunities that a business may choose to pursue or not. Having such options provides the freedom to make optimal

---

[1] BPP Learning Media, *Advanced Financial Management* (U.K.: BPP Learning Media Ltd., 2015), page 169.

decisions, such as when and where to proceed with a particular task. Various management actions, such as making strategic investments, can create real options, allowing companies to take further actions in the future based on prevailing market conditions. In short, real options are about companies making decisions and choices that grant them the greatest amount of flexibility and potential benefit regarding possible future decisions or choices.

For an option to have value, there must be uncertainty about future cash flow. An option can exist without uncertainty, but in that case, it has no value. Also, the management must have flexibility to respond to the uncertainty as it evolves. Read more about uncertainty in the section Types of Uncertainty and How They Affect Options Valuation in Chapter 7.

"Why We're Hardwired to Hate Uncertainty"

Before we move on with real options, it is worth understanding why we, human beings, hate uncertainty and why we should look for means to decrease it.[2]

We prefer to be wrong rather than uncertain. A study by the University of London published in 2016[3] found that participants' stress levels were much higher when they did not know the outcome (50 percent/50 percent) than when they were either right or . . . wrong!

The author, Archy de Berker (UCL Institute of Neurology), explained, "Our experiment allows us to draw conclusions about the effect of uncertainty on stress. It turns out that it's much worse not knowing whether you are going to get a shock than knowing you definitely will or won't. We saw exactly the same effects in our physiological measures—people sweat more and their pupils get bigger when they are more uncertain." Co-author Dr. Sven Bestmann (UCL Institute of Neurology) adds, "From an evolutionary perspective, our finding that stress responses are tuned to environmental uncertainty suggests that it may have offered some survival benefit. Appropriate stress responses might be useful for learning about uncertain, dangerous things in the environment."[4] When our ancestors heard a weird noise that they were uncertain about, their stress would rather advise them to run away than analyze whether the noise was made by a friend or a by lion trying to eat them. This response helped us survive, but today we still experience the consequences of the way we evolved. The implication here is that we are naturally conservative; as the saying goes, "Curiosity killed the cat." This proverb suggests that this behavior of being stressed by uncertainty has been recognized for centuries, if not millennia.

Even though we don't participate in the lab tests or run away from wild animals daily, uncertainty still causes us a lot of stress. Let's consider an example from our own lives:

Imagine you are driving to meet a client. The traffic is light, and you're likely to arrive on time, so there's no need to rush or worry. However, if you're stuck in heavy traffic and are highly likely to be late, you might as well relax and think about making your excuses. But if it's close—if your odds of making it on time are 50/50—that's when you will experience the most stress. Your striatum is flooded with dopamine, and its job description requires it to do something, anything, to improve those odds.[5]

The same mechanism leads to many dysfunctions in the way we manage our processes, particularly in Upstream discoveries. Whenever you are sitting on a call with clients or in a meeting with your manager and someone asks, "OK, so when will it be done?" If you don't know the answer, your brain will replace "I don't know" with more certain information, even if it's totally made up and unreliable.[6]

---

[2] https://www.theguardian.com/commentisfree/2016/apr/04/uncertainty-stressful-research-neuroscience

[3] https://www.nature.com/articles/ncomms10996

[4] https://www.sciencedaily.com/releases/2016/03/160329101037.htm

[5] See footnote 2.

[6] *Madagascar2: 69 Years?* https://www.youtube.com/watch?v=quGOlbNnWaY&t=30s

The benefits of applying real options are significant:

- Real options shift the focus away from trying to predict the future "perfectly" to identifying what can (or should) be done about responding to uncertainty.
- This emphasis on responding effectively to uncertainty fosters a discipline that extends throughout the entire project life cycle, not just at the initial decision-making stage.
- Real options enable decision makers to identify the optimal level of flexibility by providing information about the additional value that increased flexibility can offer.

As noted previously, the term *real options* refers to work that uses tangible assets. Although the Kanban Method focuses on the improvement of intangible knowledge work, there is still much to learn from how real options can be utilized and managed.

Categories of Real Options

From the financial management perspective, there are three main categories of real options: those relating to the work's size, to the project's lifetime, and to the broader business context.

Group 1: Options Relating to the Size of a Project

This group includes options to expand the project over time, given certain market circumstances and opportunities.

Group 2: Options Relating to the Lifetime of a Project

In this group, we find options to initiate work, delay (or defer) its start, redeploy resources, or abandon an existing project.

Group 3: Options in Relation to Other Work

This group includes options for process flexibility, product mix, and operating scale.

Options in Upstream Kanban

By its nature, the work in Upstream Kanban is not committed and remains *optional* until the commitment point. This means that at each step of the discovery process, the work item can be either advanced in the process, shelved, or discarded.

In that sense, Kanban's real options align with Group 2 of the financial options—those related to the timeline of the work. Let's examine them one by one.

Option to Invest

- Analysis of the work item indicates favorable outcomes.
- There is a low level of uncertainty, suggesting that this item should be prioritized for investment.

- This option can also transform into an option to expand if the analysis indicates that expanding the scope improves the Net Present Value.

> **Net Present Value** (NPV) is the cash flow expected as a result of a new project, but that flow is discounted by a rate that could otherwise be earned by doing nothing. If, for example, you would normally invest in US Treasury bonds that pay 3 percent, the project or the cash flow from it must yield a return of more than 3 percent; otherwise, it wouldn't be worth pursuing—you might as well leave the money where it is. Therefore, in this case, the end value (present value) of the project or investment needs to be discounted by 3 percent to reflect the return on the Treasury bonds.[7]

Option to Shelve

- Delays proceeding with the work item until a later date.
- Applies when there is high uncertainty of returns or not enough information about the product.
- Can be converted into an invest option when economic conditions are favorable.
- Can be discarded when economic conditions remain unfavorable.

Option to Discard

- Analysis of the work item indicates unfavorable outcomes.
- Low level of uncertainty—we know this item should not be considered for investment.
- The cost incurred needs to be considered a sunk cost.
- Although this particular item is discarded, the analysis results may be reused (re-deployed) for another project.

Real options are most appropriate when the economic environment and market conditions relating to a particular project are highly volatile yet flexible. Stable or rigid environments will not benefit much from managing options. Whenever the demand accepted by the company remains irrefutable,[8] it is impossible to introduce full options thinking, as the work item is committed from the start of the process. Then the company's choices are to use:

- The invest option with the option to delay beginning the work.
- The embedded options (to design or to implement) even when committed to a project, product, or feature, but the form the deliverables take may still be negotiable.

Also, having real options is applicable only when a firm's corporate strategy supports flexibility and there is sufficient information flow.

---

[7] https://www.investopedia.com/terms/r/realoption.asp
[8] https://www.investopedia.com/terms/a/abandonmentoption.asp

> #### *Irrefutable Demand*
>
> Irrefutable work is not discretionary. There is no option to decline it. A full triage decision is not possible. Only the *now* and *later* options exist with irrefutable work. Work is irrefutable because it is already committed, and the decision to commit to it was made elsewhere, by someone else, often at a much higher pay grade. Irrefutable demand represents a problem because it is often a source of overburdening, and hence, a root cause of disappointment and service delivery that is not fit for purpose.[9]
>
> Irrefutable demand can be for a non-optional piece or part of an already-committed whole. The need for the part is essential; otherwise, the whole will not function (or will not be of value), totally unfit for purpose.

Common Problem

There is a recurring question: "Why should I invest my time and money in discovering knowledge about items that may ultimately be discarded? If I start working on something, shouldn't I just deliver it?"

As discussed previously, we introduce options when we are facing some degree of uncertainty about the future to allow ourselves greater flexibility and a wider choice of the most fitting solution.

Time spent managing options (even if they are discarded) is an insurance cost. When you decide to buy life or health insurance or to insure a car or a home, you do it because you are *uncertain about the future.* You never know whether or when you will need to collect on the car or home insurance. Hopefully, never! But if something were to happen and you didn't have insurance, the cost of fixing the problem might exceed your individual capabilities.

This is how options work. In an uncertain environment, they provide insurance that you will avoid bearing the high cost of delivering work that doesn't fit your customer's needs or is not consumable by any of your clients. With options, you know in advance what's worth investing in and what to abandon.

Designing Real Options into the Upstream Kanban System

The optionality embedded in Upstream Kanban will mostly be reflected in the system policies, which should describe how to treat the items at each decision point prior to the commitment point.

In the iterative process used at the David J Anderson School of Management (Figure 3), each iteration concludes with a decision on whether each idea (option) should be invested in, shelved, or discarded.

Ideas for which an investment decision was made were promoted to the next column. An idea with high, yet unresolved, uncertainty was moved to the *Shelved Items* row on the board. If the decision was to cease working on the idea, it was moved down to the *Discarded Items* row.

---

[9] *Kanban Maturity Model*, 2nd Professional Coaches' edition, by David J Anderson and Teodora Bozheva (Seattle: Kanban University Press, 2021), page 209.

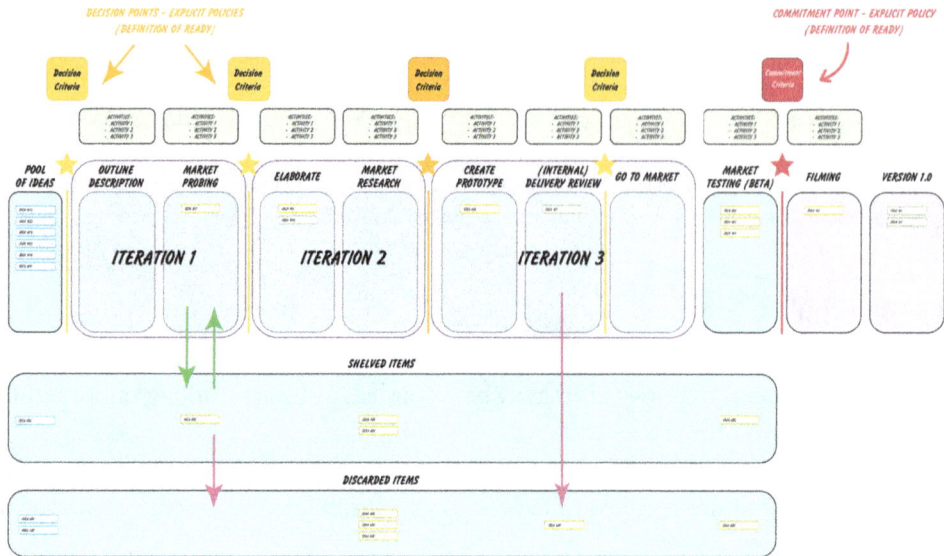

Figure 3 Visualization of the iterative process used at the David J Anderson School of Management

At each discovery process stage, the idea is either:

- In progress: The discovery process is happening; we are collecting information and decreasing uncertainty.
- Waiting for a decision: The discovery process has been completed, and we need to make an Invest-Shelve-Discard decision.
- Waiting for others: If something is blocked Upstream, we cannot acquire information. A blocked option undermines the value of an option. Although we recognize that this should not happen, this may occur in lower-maturity organizations. If you observe this situation, discuss the root causes and take action during a *Retrospective* or *Improvement Suggestion Review* to eliminate waiting items. Read more about visualizing and managing blocked items in Chapter 2, Designing Upstream Kanban.

Implementation Guidance

- At each decision point, we develop a local "definition of ready" to define when we can consider an item ready for further work; that is, it meets the investment criteria (favorable outcomes, low level of uncertainty). The investment decisions may include (but are not limited to):
  - Simple investment: moving an item farther along in the process to continue the knowledge discovery process

- Redeployment: reworking an idea into a new option
    - Expanding an idea by merging options
- Whenever investment criteria are not met, the idea cannot be promoted, and a further decision needs to be made on whether we shelve or discard it. To facilitate this, we also create an explicit policy:
    - We discard items for which we know—with a low level of uncertainty—that they will not bring us a positive, favorable business outcome.
    - We shelve items for which the uncertainty level remains high, and we don't have enough information to predict the outcome at the decision point.
- Shelved items can either go back into the process at the point where they were shelved or eventually be discarded. Until discarded, they are treated as "work in progress."
- For discarded items, the process has finished. Therefore, if any circumstances indicate that a previously discarded item should be brought back, it should be re-created and placed as a new item in the Pool of Ideas.
- The process of making Invest-Shelve-Discard decisions should be repeated every time you reach the decision point in your process.
- To keep the right cadence of decision making, establish a regular Service Request Review, or Replenishment Meeting (read more in Chapter 4 Upstream Feedback Mechanism: Meetings and Reviews).

A Supporting Case: DHS

You can learn more about implementing real options thinking from the case study[10] by Nate Conroy and Marigny Boyd, who implemented the Kanban system at the US Department of Homeland Security (DHS).

Real options were introduced at the project level, where it was required to decide which ideas should be promoted and which should be discarded. Therefore, the ideas went through the discovery process of understanding the technology, all related policies, and benefits of investing limited resources. In conclusion, the projects were either:

- committed to
- reworked into a new option
    or
- declined.

This case study demonstrates how to use real options as a means to introduce flexibility into your process and ensure that what you ultimately choose is the best option.

[10] https://www.youtube.com/watch?v=M3H0AzmET7o

Takeaways

- The concept of real options originated from the financial management sector, and in that context referred to work involving tangible assets.
- Real options are opportunities that a business may or may not take advantage of or realize.
- For an option to have value, there must be uncertainty about future cash flows. Although options can exist without uncertainty, they have no value in such cases.
- People hate uncertainty. We prefer either to be right or wrong—but not uncertain!
- In Upstream Kanban we recognize three categories of real options: invest, shelve, or discard.
- Time spent managing options (even if they are discarded) is an insurance cost.
- In an uncertain environment, real options provide insurance that prevents building or delivering something that no one needs.
- The optionality embedded in Upstream Kanban is mostly reflected in the system policies, which should describe how the items should be treated at each decision point prior to the commitment point.

Additional Resources

Reinertsen, Donald G. *The Principles of Product Development Flow: Second Generation Lean Product Development* (Redondo Beach, CA: Celeritas Publishing, 2009).

https://www.investopedia.com/terms/e/expansion-option.asp

https://www.investopedia.com/terms/o/optionpricingtheory.asp

https://www.investopedia.com/terms/o/optionscontract.asp

https://www.accaglobal.com/gb/en/student/exam-support-resources/professional-exams-study-resources/p4/technical-articles/investment-appraisal.html

2

Designing Upstream Kanban

The Upstream Kanban typically doesn't exist on its own, separate from the Downstream Kanban. What we usually observe is that Upstream Kanban evolves as a consequence of scaling Kanban width; however, this is not always the case. In this chapter, we look at the basic practices applicable to Upstream Kanban design and then explore the commonly recurring patterns:

- Two-board design
- Three-board design
- Iterative discovery

It's worth mentioning that the upstream activities can be performed by a separate organizational unit, often in a different location; hence, there are two separate boards for upstream and downstream. In other cases, it tends to be one big board, mostly operated by the same team of people, in the same (virtual) geographic location.

Upstream Kanban Design Practices

Maturity Levels 0–2

In the most natural and obvious way, the Upstream Kanban emerges as a service matures, when the question about customer demand starts to be raised. The place where those questions are answered is the Upstream Kanban, which creates the need for visualizing the process of discovering customers' needs. We observe this behavior at the transition to Maturity Level 3, and this is where we start observing most of the Upstream Kanban practices.

Day-to-day experience shows, however, a different pattern. Most companies, even those in which the services are (in general) of lower maturity,[11] are already aware of the need for a visual representation of the discovery process. Our awareness of the customer's wants and needs is usually high, even though other values may still require significant improvement.

To avoid overreaching and abandoning Kanban adoption, it's worth starting with the lower-maturity practices that will enable the company and its services to grow without overwhelming the system—which often leads to strong change resistance.

The following practices can be successfully applied at any point in the process (in the delivery service and in the emerging Upstream Kanban), even at the lower maturity levels. You can safely apply these practices when you are still discovering your process and identifying whether you operate in the Upstream area. Also, if you are a Product Owner, Business Analyst, System Analyst, or play a similar role, you can apply them to your work and the work of your team.

1. Visualize the process on the kanban board.

   - Identify the main activities that you perform from the moment when the work arrives until you give it to the delivery team.

   - Use the corresponding verbs that describe the essence of the activities. Avoid the temptation to state which role (who) is responsible for doing a job; instead, focus on what must be done in a particular state.

   - The Upstream part of the workflow represents the steps through which an idea evolves and converts into a committed request; for example, *Business Proposal—Technical Approval—Ready for Development*.

   - You don't know exactly what happened? Create a simple kanban board with *Pool of Ideas—To Do—In Progress—Ready for Commitment/ Development/To Pull*.

   - **Important: Do not copy existing solutions!**
   There is a recurring question students bring back in our classes: If I don't know what my upstream process looks like, should I use the design of the other team (which seems similar) and later adjust it?

     No. You should instead start with a generic design (*Pool of Ideas—To Do—In Progress—Ready for Commitment/ Development/To Pull*) and let the upstream process emerge from the *In Progress* column. The risk in copying an existing solution is getting stuck and trying to fit into someone else's process, which may differ from yours. By letting Upstream emerge, you make sure that your visualization represents the actual flow of *your* work.

---

[11] Maturity levels (referred to as ML1 through ML6) are described in *Kanban Maturity Model*, 2nd Professional Coaches' edition, by David J Anderson and Teodora Bozheva (Seattle: Kanban University Press, 2021). For the basic practices: https://all.kanban.plus/en/content/kanbanplus/kmm/default/kmm-posters/poster/practice-map-poster

2. Position the cards that correspond to each work item in the column that represents that item's stage of work progress.

3. Visualize WIP limits on active (*In Progress*) and committed (*To Do*) work. At the lower maturity levels, you can start with per-person WIP limits.

4. Use avatars to visualize who is doing what—the current team focus. Each individual may have more than one avatar. The number of avatars per person usually corresponds to an agreed per-person WIP limit.

Supporting Case: BestDay

An example of starting simple can be found in the BestDay case study by Alex Rodriguez.[12] To avoid previous issues with overreaching and designing a "perfect solution" up front, their Upstream process consisted of only two activities: *Discover Quarterly Objectives* and *Define*.

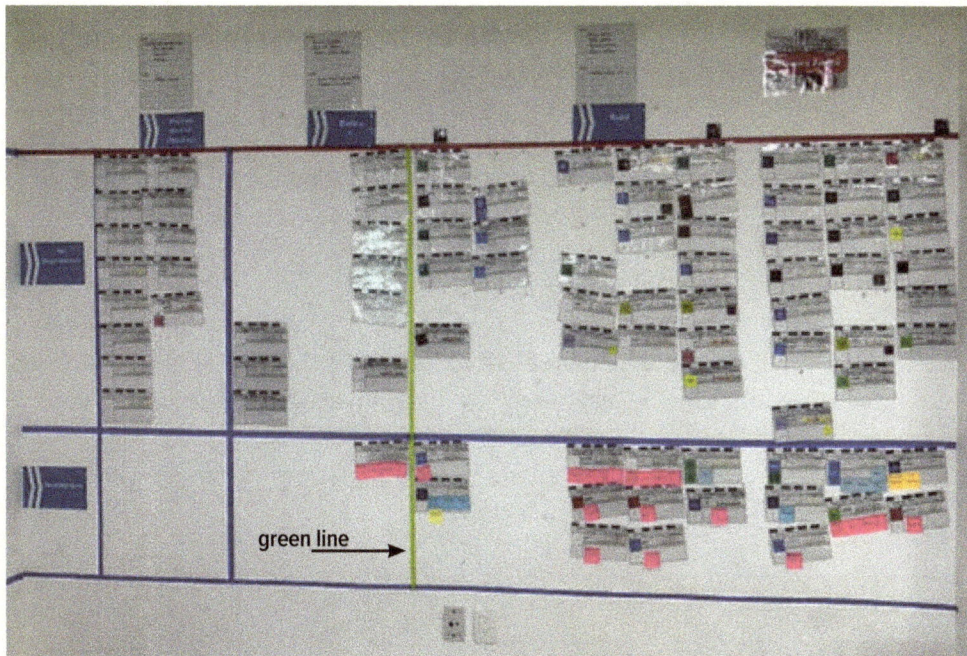

Figure 4 Process visualization at BestDay with Upstream and Downstream systems, with rows indicating dependencies

The vertical green line represents the commitment point, and the rows on the board indicate dependency/no dependency type of work.

The problem of big items stuck in the generic Upstream process was resolved by a proper breakdown of work items into smaller—but still bringing real value—pieces of work.

---

5. Visualize work types using card colors or board rows:

- Establish a color scheme for visualizing work item types.
- Alternatively, use separate rows on the kanban board to visualize different work types. If work items are categorized by size, different-sized work items can be represented on physical kanban boards using different-sized cards.

6. Visualize work item aging.

7. Visualize blocked items (Figure 5). Create a visual indicator on the ticket or a parking lot for items that are waiting for someone else to take action or provide information.

> **Note:** The parking lot is not a trash bin! Don't let items be forgotten.

- Visualize basic information and establish policies that will help minimize the time wasted by work items in the "waiting" state.

  Create a blocker ticket that indicates:

  ◦ When the blocker started.
  ◦ When the blocker ends.
  ◦ Why is it blocked? Or where is it waiting?
  ◦ When would you need to follow up? Agree on the follow-up date with those responsible for providing you with information or the solution in case no update arrives.
  ◦ Who on the team is responsible for this blocker?

  Create explicit policy:

  ◦ How often do you review the blockers? Don't let them rot in the parking lot.
  ◦ What actions can you take when you are not receiving the required information or solution? Where can you look for help?

  Collect blocker-related information for *Team Retrospective/Flow Review* or (at deeper maturity levels) *Blocker Clustering*.

IDEA ABC

Blocked on: 13 Jan 2023
Unblocked on:
Waiting for: Risk analysis
Next follow up: 17 Jan 2023
Responsible: Anna

Figure 5 Example of a blocker ticket

8. Visualize dependencies on shared services using avatars.[13]
9. Identify work types based on customer requests, taking into consideration the following aspects of the work:
   - Source of demand: the types of customers who may request work; for example, internal or external; the types of end users
   - Customer requests (even if unreasonable): for example, receive information, resolve an incident, obtain a particular report, get a new feature for a software application, and so on
   - Expected outcome of the work: for example, the type of product or service
   - Specialist skills required: for example, development, marketing, or commercial
10. Collect flow-related data (read more in Chapter 5, Upstream Metrics).
11. Establish a feedback mechanism (read more in Chapter 4, Upstream Feedback Mechanism: Meetings and Reviews).

Upstream Kanban in Fit-for-Purpose Organization

In more mature, fit-for-purpose organizations[14] (Maturity Level 3 and deeper) Upstream Kanban is defined, and the commitment point is no longer ambiguous. Instead, the discussion about commitment is a vital moment in the end-to-end process when the work loses the "optionality" feature.

An Upstream kanban board is intended to facilitate the development of a stream of ideas before they are converted into committed work. The Upstream part of the workflow represents the steps through which an idea evolves and becomes a committed request. Instead of handling a single, unordered list of customer requests, the discovery kanban board allows the team to focus their efforts on elaborating options, concentrating first on those that can be replenished to the delivery team while further elaborating alternatives.

Elaborating options often involves individuals who also work Downstream, that is, those who also have assignments on the delivery kanban board. For instance, developers might work on a prototype or a proof of concept Upstream and on implementing functionality Downstream.

Creating a smooth flow of options without affecting the delivery workflow requires proper management of each individual's availability.

Implementation Guidance

- Use an Upstream kanban board to visualize the state of possible options. The commitment point separates the discovery kanban board from the delivery board. Only requests that the customer is sure they want delivered pass through the border between the two systems.

---

[13] D. J Anderson and T. Bozheva (2021), p. 131.
[14] D. J Anderson and T. Bozheva (2021), p. 141.

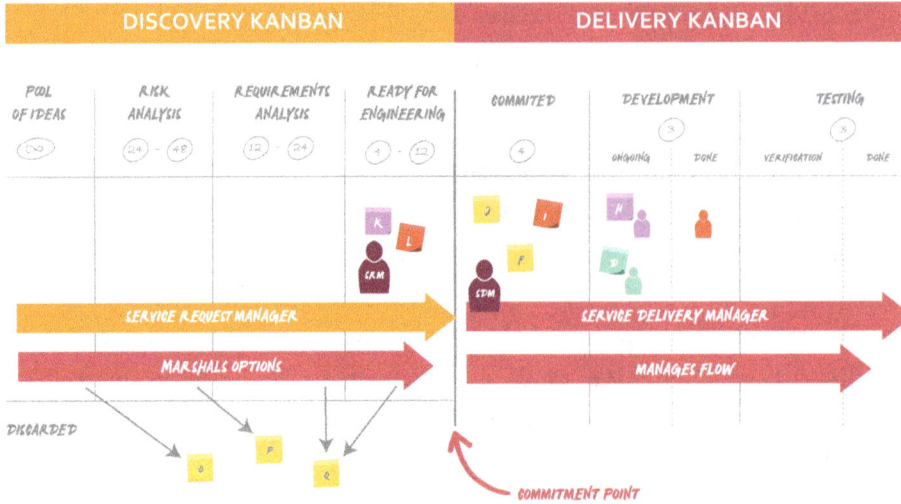

Figure 6 Example of a kanban board visualizing Upstream and Downstream work with a commitment point and discarded items

- Visualize the steps through which a request or idea evolves through analysis and synthesis (Figure 6); for example, *Pool of Ideas—Risk Analysis—Requirements Analysis—Ready for Engineering*. In the *Pool of Ideas* stage, an idea is rather rough and unspecified. In the *Requirements Analysis* stage, the idea becomes coherent, and clearer, and can be split into concrete smaller requests. In the *Ready for Engineering* stage, the requests that have been defined from a business point of view are validated from a technical perspective. Once this approval is obtained, a request is committed and is ready to be pulled into the Downstream (Delivery) Kanban system.
- Use a ticket to visualize an idea (option).
- Use avatars to visualize the workload of individuals working Upstream.
- If the individuals working Upstream also work Downstream, extend the per-person WIP limits to include the work items from both the Upstream and Downstream kanban boards.
- Often, discovery and delivery are performed by separate organizations, and in many larger enterprises these functions do not sit together. When physical boards are used, it is common for a discovery board to be separated from a delivery board; it is more natural to have two boards rather than just one.

Examples of Upstream Kanban Design

Three-Board Design

In a three-board design, the workflow is designed as follows:

- Board 1 becomes an Upstream board ("what" is decided).

- Board 2 becomes the middle ground between client and vendor, Product team and Development team, customer and delivery, and so on ("how" is discovered).
- Board 3 becomes a Downstream board (solution delivery).

Supporting Case: Nemetschek SCIA

The first documented case of using Upstream Kanban as an element in a three-board solution was Patrick Steyaert's Nemetschek SCIA case study (2009–2013).[15] Their implementation (shown in Figure 7) consisted of:

- **The Discovery kanban board:** Ideas were created and developed by the team; Product Managers picked ideas from the backlog and put them on the board; after Product Managers acquired more information, they decided on which items would eventually make it to the second board and which would not.
- **The Expert System Requirement kanban board:** When the Product Managers recognized a work request as a vital component, it was moved to the second board, which created the possibility for developers to influence the work item's destiny; this Expert board was a venue to hold a discussion between Product Managers and Developers and decide together the future of the work and, eventually, also agree on the requirements so that the developers would stick with it during the follow-up delivery.
- **The Delivery kanban board:** This was a regular, Downstream board where the bigger requirements from the second board were sliced into smaller pieces of work that developers pulled whenever capacity was available.

The Expert board was designed to provide a platform for conversation between Product Managers and Developers and to restore the Developers' ability to influence a product's future. In the Nemetschek case, the middle board helped with filling the previously created gap between the two parts of the end-to-end system.

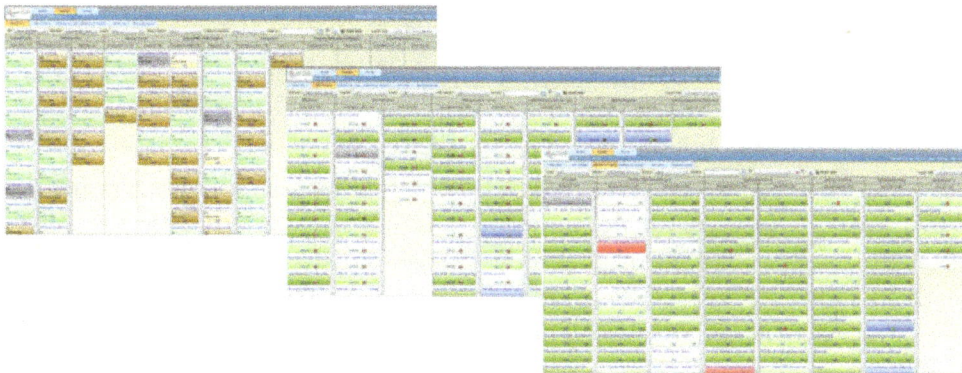

Figure 7 Screenshot from SCIA Engineering's three-board solution

---

[15] https://all.kanban.plus/en/content/kanbanplus/bwk/default/bwk-case-studies/reader/scia-kc

Supporting Case: FARA

I was running the *Kanban System Design* class with a group of Development managers, Product Owners, and Test managers. We reached the point in the class where we started applying STATIK[16] (the Systems Thinking Approach to Implementing Kanban) for their service. One of the sources of dissatisfaction was frequent communication issues between the Product and Delivery units. The Product team was unhappy with Development and Testing not delivering on time, which caused constant tensions and problems related to customer contracts. The Delivery team complained that the requirements provided by the Product Owners were "not good enough," and that eventually each piece of work became a much bigger effort than originally planned. They were dealing with an enormous amount of "dark matter" (read more in Chapter 7, The Value of Deferred Commitment).

Figure 8 shows how the process looked back then.

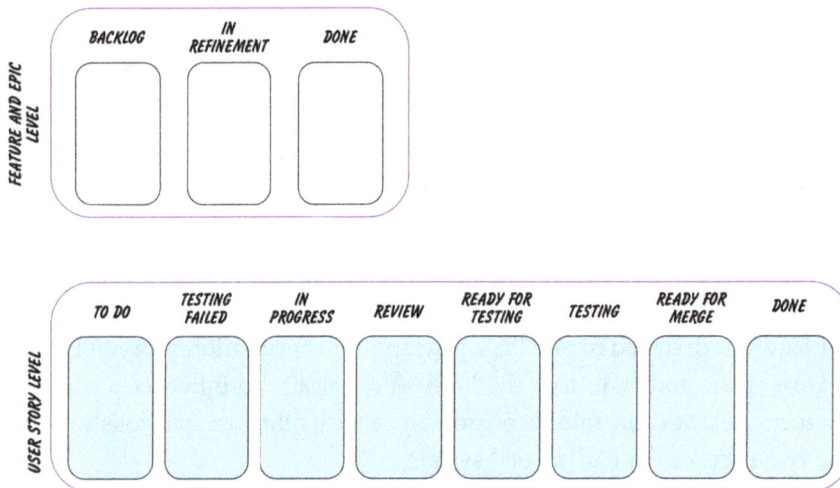

Figure 8 Work visualization in FARA at the kickoff of the STATIK workshop

There was no way to see the whole process; items under *Done* from the *Feature and Epic* level were pushed to Development's *To Do* and planned for the sprint. There was almost no discussion about product features and requirements.

Going through the STATIK steps helped everyone realize what they were facing and what they needed to focus on to solve the issue. Figure 9 shows the result; however, sometimes the beautifully designed kanban board is the least important part of the process.

Starting at the top, the upper board represents the end-to-end process from the moment a client requirement is translated from the contract agreement into the work item (feature) to when it is delivered. The client paid for the delivered feature. A *Feature* could consist of a few *Epics*, which were handled by Product Owners.

---

[16] https://all.kanban.plus/en/content/kanbanplus/bwk/default/bwk-articles/reader/statik-article

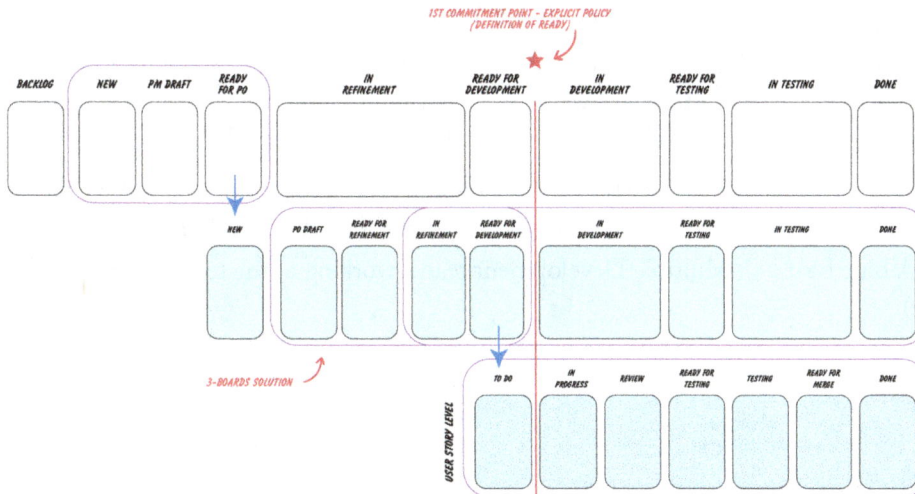

Figure 9 Workflow visualization as the result of the STATIK workshop

The middle board represents the three-board solution like the one at Nemetschek SCIA:

- *PO Draft* and *Ready for Refinement* cover Product Owners working with clients and Product Managers to explore and understand the clients' needs.

- *In Refinement* and *Ready for Development* represent the platform where Development teams were invited to discuss the requirements and refine the *User Stories* created by Product Owners for their *Epics*.

- *The Definition of Ready* (the Acceptance Criteria policy) at the *Commitment Point* was established to ensure that no work items are pushed to the delivery board until the scope of work and available capacity are understood.

Acceptance Criteria Policy

In the *Kanban Maturity Model*, the practice XP 3.2, "Explicitly define request acceptance criteria,"[17] is crucial. Establishing explicit policies for accepting Upstream requests facilitates the process of deciding which options to pull into the *Ready-to-Start* stage of the delivery kanban system and when that can happen.

Implementation Guidance

Decide what information the customer request must include to initiate the delivery process. The following criteria are typically used:

- ▸ It is complete.
- ▸ It is clear.
- ▸ It is coherent with other requests.
- ▸ It is testable.
- ▸ It has defined user-acceptance criteria.

Make sure that enough capacity is available to commit to delivering the work item within the expected time.

---

[17] D. J Anderson and T. Bozheva (2021), p. 238.

The bottom board represents the delivery part of the system and is owned by the Development team. Actually, there were a few boards like this, all fed by the middle (epic) board. The work was committed and went through the *Definition of Ready*.

Figure 10 depicts the vertical structure of product building at FARA:

- White board(s): Product Management working at the aggregated level (client features)
- Light blue board(s): Product Owners working at the middle level of epics
- Dark blue board(s): Multiple Development teams working at the team (user stories) level

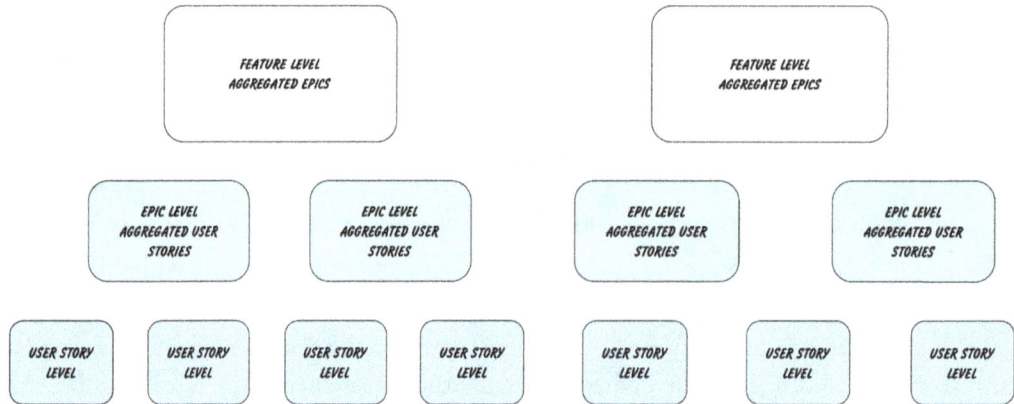

Figure 10 Teams structure at FARA

Supporting Case: Optimizely

Optimizely[18] is an interesting case, which, at first glance, looks like a two-board design (upstream and downstream) visualized on one huge whiteboard (Figure 11). However, a closer examination of the design also reveals a three-board solution with a middle board.

The solid line in the middle shows the commitment point between the Upstream and the Downstream systems.

Let's look more closely at the Upstream, to the left of the commitment point.

It is divided into two parts:

- Problems (Research [to the left of the dotted line])—understanding the problem:
  - Iteration 1: Get data about the problem that will help you to understand it and generate ideas—without refining them yet.
  - Iteration 2: Not everything you learned is going to be useful and result in solutions, so sift them down to the most important ones, collecting more data about the ideas that you will try to use.

---

[18] https://www.youtube.com/watch?v=y8Ns5bdg0oo&t=192s

- This part of the board represents the upstream work done before the first triage question: Should we do it or not? That's the question about the "what," when the demand is still refutable and the options are the customer's options.
- Solutions (Design and Technical Feasibility [between the dotted and solid lines])—looking for solutions:
  - Iteration 1: Find as many different solutions as you can to determine which will work best.
  - Iteration 2: Refine the solutions generated in Iteration 1 for the best solution to the problem, which involves prototyping, mock-ups, sketches, and user testing of the prototypes.
  - This part of the board represents the work done between the first triage question (Should we do it or not?) and the second one (When should we do it?). That's the question about "how," when the demand coming from the customer becomes irrefutable, but the internal optionality still exists while we are discovering the solutions.

Read more in Chapter 9, Plan Your Work! Cost of Delay, Triage, and Two-Phase Commitment.

The outcome of this process is a specific solution that you can build and deliver to the customer (Figure 12).

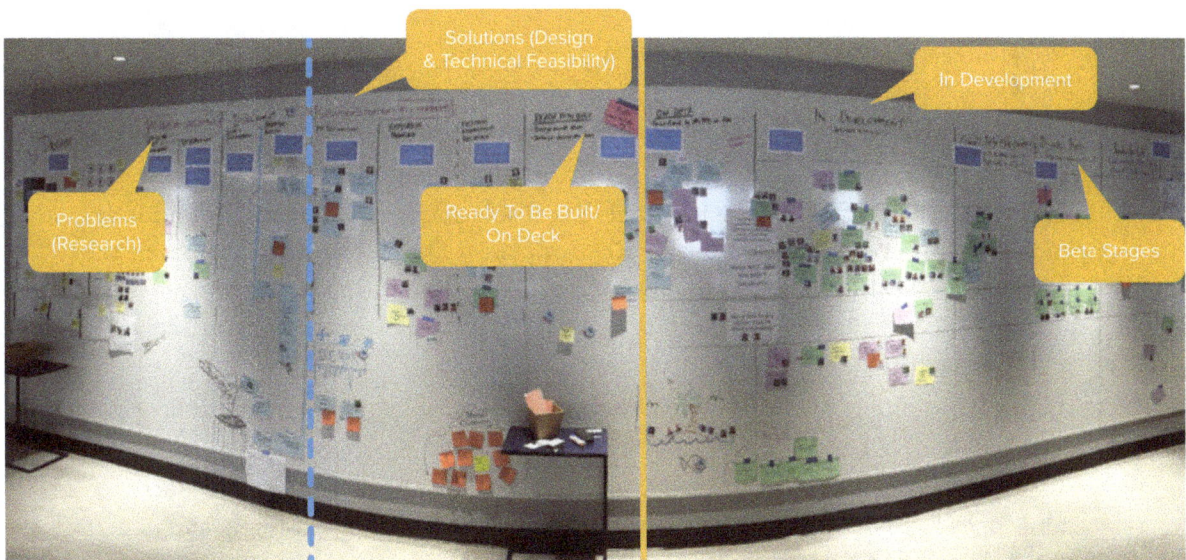

Figure 11 End-to-end workflow visualization at Optimizely

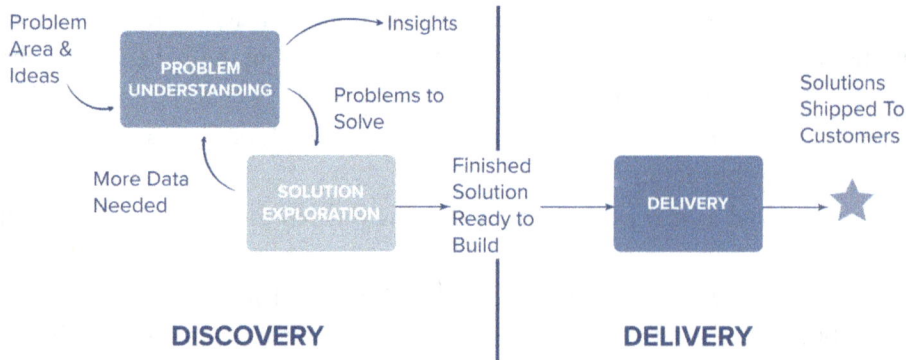

Figure 12 Diagram representing the flow of work items in the Optimizely process

There is an interesting difference between SCIA and Optimizely in that Optimizely uses the iteration between a problem/purpose/goal and a solution (a "what"), and the delivery part of their Downstream probably includes *Design*, *Build*, and *Test*. They are not that explicit about design options (exploring the "how"), while the SCIA flow is much more explicit about exploring the "how." It can be argued that the Optimizely board extends further Upstream into problem areas and ideas that need to be explored, and there is a lot of uncertainty. This may not have been the case with SCIA because the domain of civil engineering was well established, so there was little uncertainty about the problems to be solved, the goals, or the purposes to be fulfilled. The practice of civil engineering across a range of technologies and materials was already well understood. Therefore, they could explore tradeoffs and priorities of "what," but they were not exploring "why" for a purpose- or a goal-directed user base.

Arguably, Optimizely was exploring things like Who are our users? or *Could* they be our users? while SCIA already knew this. Optimizely needed to explore and test who and why scenarios, whereas SCIA needed only to test the specific details of what.

Recurring Pattern

These are the three cases where the three-board solution emerged naturally and independently. However, teaching a lot of *Kanban System Design* classes and running the STATIK sessions, I had a chance to observe this pattern appearing almost whenever the company was struggling with low-maturity processes, where the work happened in silos.

The main dissatisfactions were:

- Lack of a proper discovery system that would support understanding the client's needs.
- Partial communication or lack of communication between discovery ("product people") and delivery ("development people"), involving blamestorming and finger pointing when delivery was not on time or as expected.

The three-board solution seems to address these problems properly by:

- Engaging everyone in understanding and designing the Upstream workflow.
- Providing a platform for communication and properly deliberating the customer's needs and product requirements.

There are two stages of implementing this design:

1. In lower-maturity organizations with emerging upstream processes, commitment is normally asynchronous. Because the first triage decision (what we do next) has already been made, the demand is irrefutable, so there is no discard option (you can either invest or shelve). As the minimum WIP limits have not been introduced, the process tends to be linear rather than funnel-like, even though we consider the work to be a part of the upstream process.

2. The design can evolve into a full Upstream Kanban implementation with synchronous commitment, refutable demand (the option to discard exists), and complete triage (Should we do it or not? and When should we do it?). Read more about triage in Chapter 9.

The three-board solution enables incremental commitment. The most recent work of Barry Boehm is *The Incremental Commitment Spiral Model*,[19] and it builds on his earlier work on iterative solution development called the Spiral Model, which predated Agile by a couple of decades. So, incremental commitment has been a part of the more formal, large-scale, system-engineering program management world for at least fifteen years.

Two-Board Design

The two-board design doesn't necessarily mean two separate boards. It just indicates that the Upstream and the Downstream parts are clearly defined, and that the commitment point—with the Acceptance Criteria policy—is transparent and well-understood by all. Two boards might be common when the Upstream and Downstream people are located in separate offices. One board is more likely when the teams are co-located and is even more likely if personnel are shared across Upstream and Downstream activities. Software tracking tools make it easier to have a single board, but very few of these products allow the commitment point to be set flexibly mid-workflow, which means that two boards may be common because restrictions in the tools require that the commitment point is always at the beginning of the workflow.

Supporting Case: Investment Bank

In this case, the Kanban Method was used from the beginning of the project work. The board was designed to cover the end-to-end process, but it lacked a clear distinction between the Upstream and the Downstream portions, and it had no established commitment point. The visualization of the process, depicted in Figure 13, shows how the team-level board looked when the workflow was made transparent for the first time.

This process resulted in poor communication with business stakeholders, delivery misaligned with their expectations, and a growing inventory of failure demand (as the consequence of either undetected defects or change requests for delivered items).

---

[19] https://learning.acm.org/techtalks/icsm

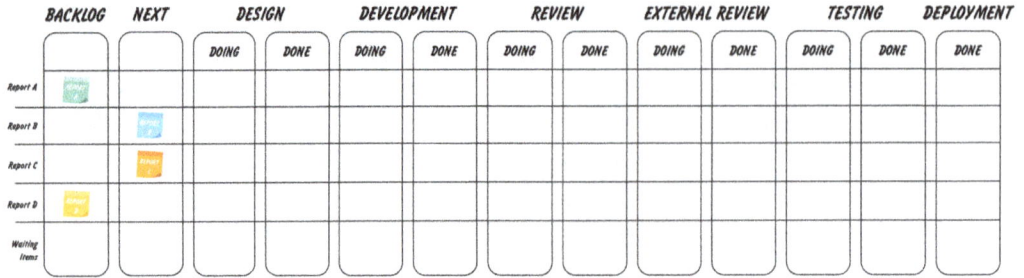

Figure 13 The first kanban board visualizing an initial state of work

To improve the situation, the process was redesigned; and it evolved, first, by:
- creating a space for the Upstream-like options analysis,
- then adding the *Business Review* column and approval for the analysts' solution proposals,
and, finally,
- creating the proto-commitment point (internal to the team, with initial acceptance criteria and a Product Owner decision about whether the feature should be developed further or not). This is depicted in Figure 14.

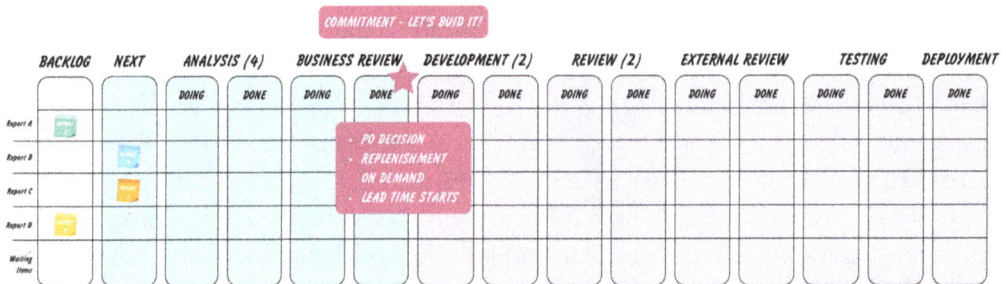

Figure 14 The kanban board at investment bank with an initial commitment point and acceptance criteria

This new design partially addressed the communication problem with stakeholders and the subsequent failure demand. However, still, the decisions were not made at the point in the process where they should have been—Upstream by the Business Owners rather than Downstream by the Product Owners (whose role at this stage was limited to being the proxy between the Business and Development teams, with no real decision-making authority).

The final design involved many more Business representatives, who supported the process of clarifying the options and making decisions on what to do next, while the Development team (including the Product Owner) provided the feedback on when they could build it. Read more in Chapter 9, Plan Your Work! Cost of Delay, Triage, and Two-Phase Commitment.

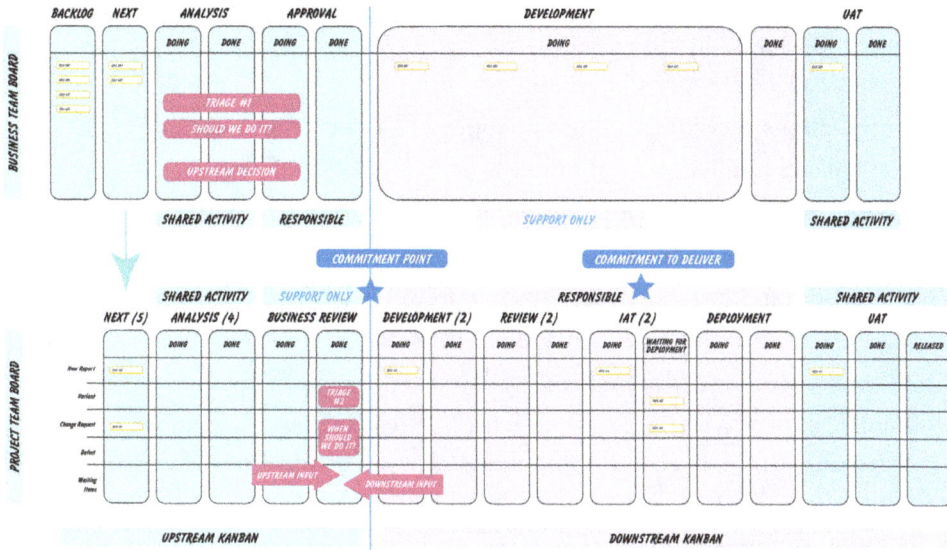

Figure 15 Process visualization with triage, commitment point, and the Business team board (Ideation group)

The following elements of the Upstream Kanban system design were applied:

- Visualization (Figure 15):
  - The ideas development process was covered by *Analysis* and *Business Review*.
  - The column *Business Review Done* indicated that items were ready to pull.
  - Replenishment signals were visualized by having clear WIP limits and *Done* sub-columns.
  - Commitment point was visualized.
  - Failure demand (defects and change requests) was visualized, tracked, and measured.
- Limit WIP:
  - Activity-based work-in-progress limits were established.
- Manage Flow:
  - The board represented the knowledge discovery process.
  - Commitment was deferred.
  - Classes of service were used to support the selection process.
  - Failure demand was measured, analyzed, and used for establishing improvement actions.
  - Triage discipline was applied.
  - Two-phase commit was applied.
  - SLA was agreed with the Business stakeholders.

- Forecasting *When will it be done?* was based on historical data and delivery cycles on the Business side (read more in Chapter 9, Plan Your Work! Cost of Delay, Triage, and Two-Phase Commitment).
- Real Options decision-making was applied in addition to triage.
- Make Policies Explicit:
  - Commitment Point was established, and the Definition of Ready was clarified.
  - Pull criteria were defined.
  - Customer expectations for each work item were clear and transparent.
  - Classes of service were defined and put in place.
- Implement Feedback Mechanisms:
  - Replenishment Meeting, merged with Service Request Review, was conducted together with the Business stakeholders (now called Ideation Group Planning).
  - Service Risk Review, merged with Delivery Planning, was conducted (now called Steering Committee).

The result was faster, smoother, and predictable delivery with improved Business and team satisfaction.

Supporting Case: Department of Homeland Security

The simple Upstream system shown in Figure 16 was created for defects management at the Department of Homeland Security. Defects in the pool were not committed; they needed either to be accepted or rejected; and visualizing the decision-making process was essential.

This flow represents the left-hand part of the defect-solving process, where items are analyzed and discussed before they are committed to by the Development team.

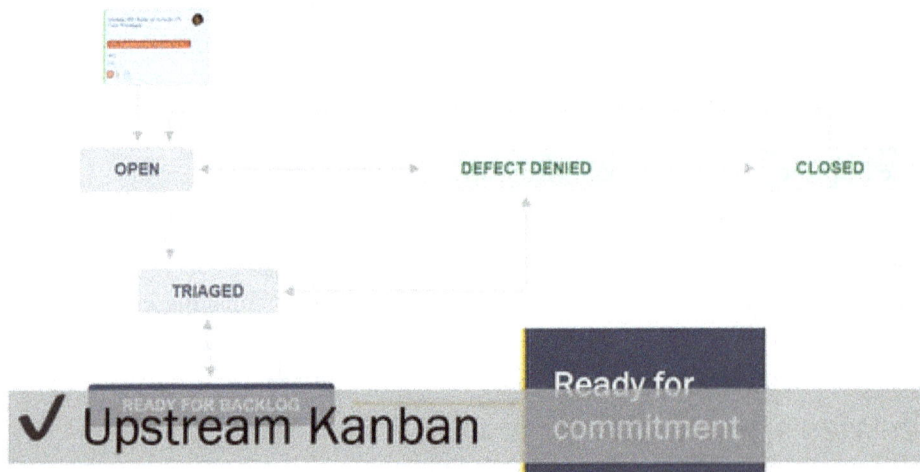

Figure 16 The flow of work items at DHS

Recurring Pattern

The two-board design resulted from an emerging Upstream process, scaling Kanban in width as the process matured. When the Maturity Level 1 process starts growing, the customer awareness of the Delivery team(s) expands, and an Upstream Kanban system emerges as a natural consequence of the needs of both Downstream systems and clients.

As we observed with the three-board design, in lower-maturity organizations with emerging Upstream processes, commitment is normally asynchronous. Because the first triage decision (Should we do it or not?) has already been made, the demand is irrefutable, so there is no discard option (you can either invest or shelve). As the minimum WIP limits have not yet been introduced, the process tends to be linear rather than funnel-like, even though we consider the work to be a part of the upstream process.

The design can evolve into a full Upstream implementation with synchronous commitment, refutable demand (the option to discard exists), and complete triage (Should we do it or not? and When should we do it?).

Usually, the two-board solution evolves either from Kanban or from other previously used methods or frameworks. Very often we observe this pattern when Kanban is implemented in an organization using the waterfall approach, a sequential SDLC process, or project-management frameworks.

Read more in Chapter 9, Plan Your Work! Cost of Delay, Triage, and Two-Phase Commitment.

Iterative Discovery

Multiple takes on collecting more information for generated options lie at the heart of Upstream Kanban, but usually, we visualize the work linearly, and the general advice says we should avoid moving items backward on the kanban board.[20] So, the question that comes back very often is how to find a compromise between the iterative nature of the Upstream processes and the linearity of a traditional design (including moving tickets on the board to the right rather than back and forth).

One way to manage this is by considering the Upstream system as a funnel, where some options are discarded before the commitment point (read more in Chapter 1, Introduction to Real Options). The discard or invest decision is the result of iterations of a discovery process. After each iteration, the number of options in the system is reduced until some reach the commitment point, where they are pulled by the Downstream system.

Also, even if the work in the iterations is repetitive, it's not identical. In each iteration, we collect a different set of data, which later informs the decision about what real option should be applied (invest/shelve/discard). By understanding the flow of information gathering and what a particular iteration covers, we can describe them and create specific exit/decision criteria. Hence, there are two layers of "gating" in the Upstream system:

---

[20] https://djaa.com/kanban-evergreen-dont-move-items-backward-on-the-kanban-board/

- Activities are visualized in columns.
- Iterations are visualized as groups of activities divided by decision points and decreasing work-in-progress limits.

That concept is depicted in Figure 17.

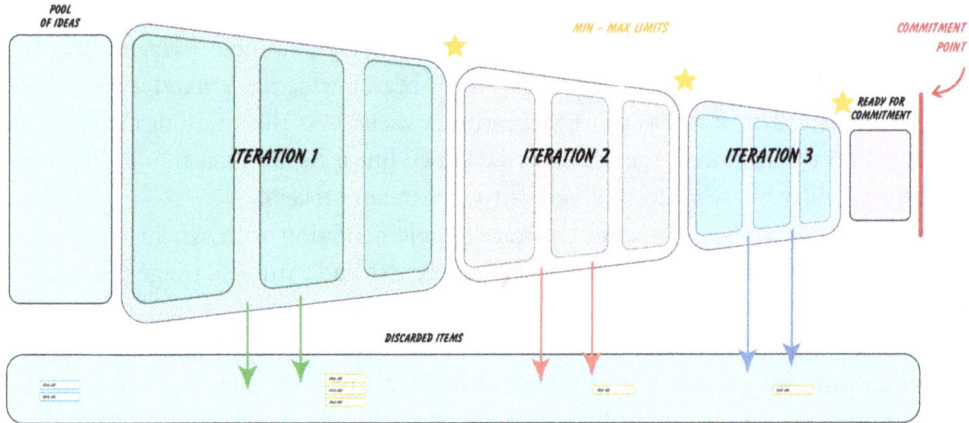

Figure 17 An example visualization of an iterative process with decision points and a commitment point

Supporting Case: David J Anderson School of Management

We designed the Upstream Kanban system at the David J Anderson School of Management to visualize the iterative nature of a class design. That design pattern is presented in Figure 18, and Figure 19 provides a closer look.

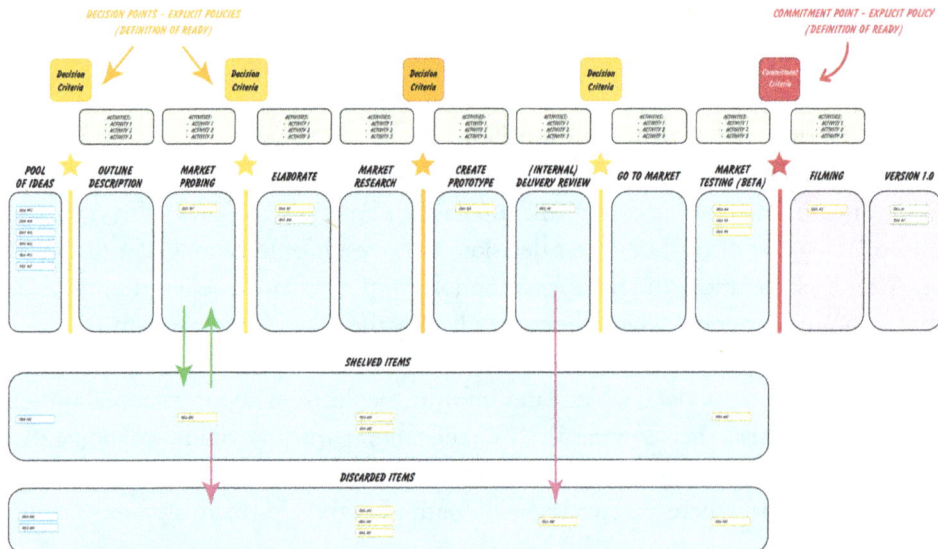

Figure 18 Iterative process for creating new classes at the David J Anderson School of Management

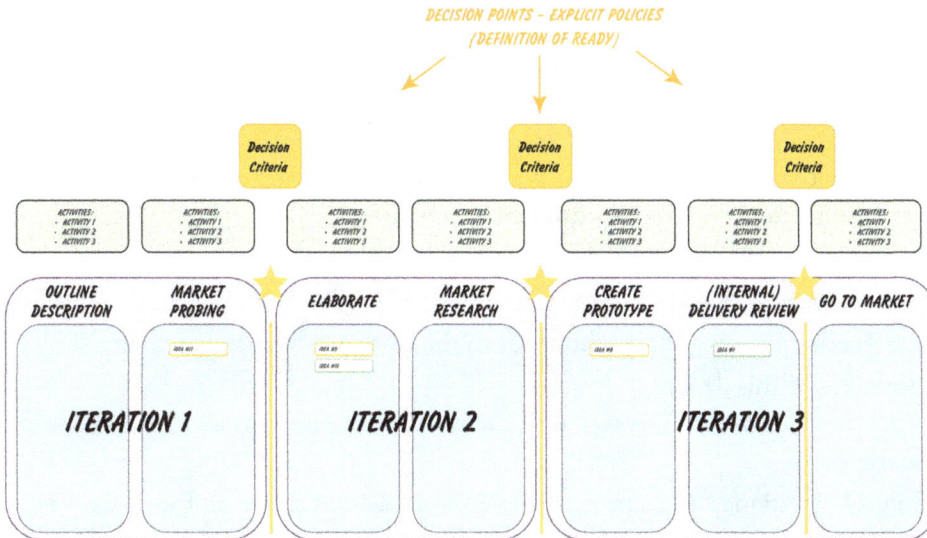

Figure 19 Visualization of three main iterations of discovery work at the David J Anderson School of Management

Whenever a new idea for class research and development is pulled into the workflow, it has gone through the following iterations:

Iteration 1:

- *Outline Description*: title, subtitle, market segment(s), brief description
- *Market Probing*: having the product description, we can send it to potential customers.

    Decision Point 1:

    - Did we hear anything back?
    - Is there any interest in the class content?

Iteration 2:

- *Elaborate*: objectives, class duration, type of the class (in-person/online), price point, where will content come from?
- *Market Research*:
    - Is there a market for it?
    - What channel will we use?
    - Do we know how to reach out to this market and find customers?
    - Can we communicate with them?

    Decision Point 2:

    - Is there enough evidence from market research to support the idea?

Iteration 3:

- *Create Prototype*: course description, key course messaging ideas, course slide deck, Miro template/exercises
- *Internal Delivery Review*: course slides edited, course slides reviewed and understood by the instructor and at least one third party, course practice run-through completion, positive feedback from market study

  Small Decision Point:
  - Are we ready with materials to start selling the product?
- *Go to Market:* planning for reaching out to the market, advertising, selling

Key elements of this design:

- Each iteration consists of two major steps: creating content and researching the market.
- In each iteration, actions become more detailed and precise, and we collect more information.
- Iterations are divided by decision points.
- After each iteration, the work item can be either discarded, shelved, or moved for further investment and elaboration.
- Each activity in the iteration has a description and exit criteria that support discussion and a decision on which real option should be applied.

This design requires deeper maturity of the service, as it implements the full concept of the Upstream funnel with multiple decision points, (ideally) synchronous commitment, complete triage (Should we do it or not? and When should we do it?), and a full set of real options. Hence, we do not observe it at lower maturity levels.

Implementation Guidance

Identify the main activities in the Upstream process.

- Identify which activities require reiteration.
- Analyze and understand the nature of these activities. Precise and describe them.
- Do all options go from one iteration to the next? When do you know whether to discard the option or move it forward? Define decision criteria for decision points.

Avoid overdesigning.

This design works well for work that requires iterating the same idea a few times over, in which each iteration adds more detail. It can be applied in processes like drafting marketing campaigns, preparing teaching content, designing visual assets, and so on.

If your process is in the form of sequential activities, the traditional, linear design is a fitter solution.

Visualization Variant: Upstream Workers Operate as a Shared Service

Big organizations very often use shared services, with some Downstream people also delivering work in the Upstream area; for example, UX designers, business analysts, and risk analysts. They function at the same time at two levels:

- As part of a shared service (e.g., the Business Analysts team)
- As members of the end-to-end process for which they deliver work

Figure 20 depicts an example visualization of that setup. The upper board represents the team of Business Analysts, who are distributed among different projects based on current business needs. The bottom board visualizes the work for project XYZ.

THE BOARD OF A PROJECT "XYZ"

Figure 20 Visualization variant where upstream workers are operating as a shared service

Let's take a closer look at each of them.

Team of Business Analysts (BAs)

The team is involved in work on four different projects, which are visualized as rows of the board. Each project represents just one swim lane on the BAs' team board, which tracks

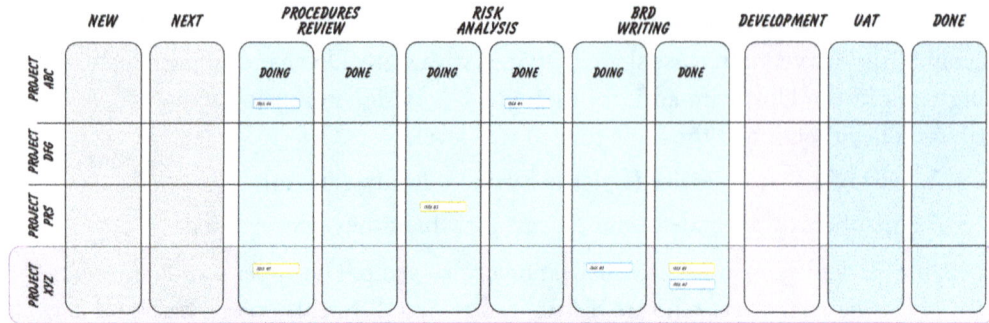

Figure 21 The Business Analysts' team board

the work for multiple projects in which BAs are engaged as shared service. Two of those Business Analysts ("Blue BA" and "Yellow BA") are working on three projects in total.

Procedures Review, *Risk Analysis*, and *BRD Writing* are considered activities of the *Analysis* columns on the upper board in Figure 20, which is the equivalent of an Upstream system. From the visualization in Figure 21, we can read that Blue BA and Yellow BA are shared, non-instantly available services. In addition to project XYZ, they are also involved in projects ABC and PRS, respectively. Those constraints should be considered when setting WIP limits.

Development and *UAT* represent the Downstream part of the process and are still visualized on the BA board as part of the end-to-end project visualization and as activities that may potentially require help from BAs (developers' inquiries, UAT support).

Project Board

The *Analysis* column represents the Business Analysts' work. On the project board, this work is aggregated as one activity; the team board shows the sequential steps that comprise that activity. Of course, the project board can also include these steps, but that was unneeded in this case.

Swim lanes are utilized on the board to visualize the flow of each type of work item identified in the project. The WIP limit on the *Next* column recognizes that BAs are a non-instantly available shared service, so there is likely some delay in the availability of analysts; hence, a buffer is required to ensure that Upstream flow continues until analysts become available.

Bottlenecks and Non-Instant Availability

Non-instant availability[21] tends to be a problem with shared resources—such as expensive special-purpose machines, systems environments, or people who are asked to perform multiple functions—that are time-sliced among different sources of demand. As we all know, there is no such thing as multitasking in the office; what we do is frequent task

---

[21] https://all.kanban.plus/en/content/kanbanplus/kbb/default/kbb-chapters/reader/kbb-ch17

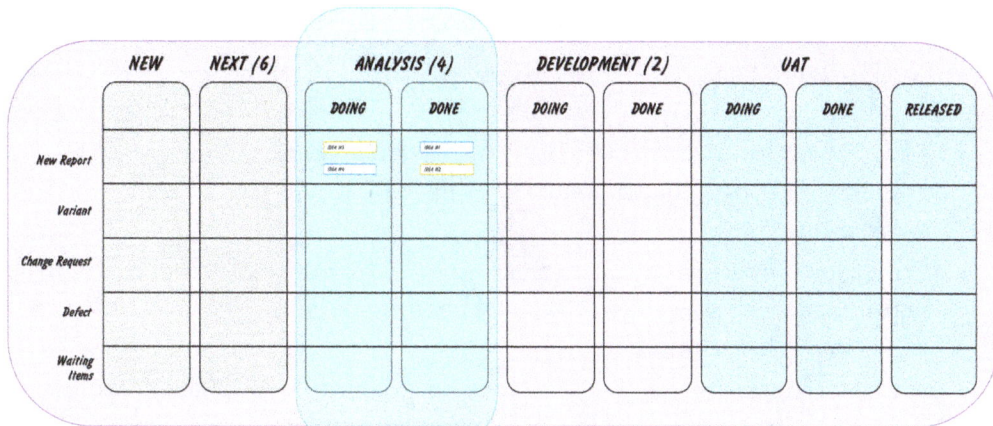

THE BOARD OF A PROJECT "XYZ"

Figure 22 The project for which Business Analysts serve as a shared service

switching. At any given time, the Business Analysts are asked to work on two or more projects simultaneously. And even though, from the project team's perspective, they focus their efforts on Project XYZ only, in reality, BAs share their time with two other projects and so can be perceived as non-instantly available resources.

One technique that addresses the problem of non-instant availability is to make sure that the work always flows in the Upstream system even when shared services are not instantly available. It is important that there is adequate capacity in the non-instant availability of a shared service to ensure that it does not become a bottleneck—a capacity-constrained resource. The buffer size should be large enough to ensure Upstream flow, but small enough so that it can be emptied each time the shared service becomes available. Hence, we put a buffer of work in front of *Analysis* to ensure that flow continues smoothly despite the non-instant availability of the analysts (Figure 22).

Takeaways

- The Upstream Kanban emerges as a consequence of a maturing service when questions about customer demand start to be raised.
- At lower maturity levels (ML0–ML2), organizations already are aware of their customers' needs, even though other values may still require significant improvement.
- At those lower levels, an organization can start with practices that will enable the company and its services to grow and will not overwhelm the system, which can lead to strong change resistance.
- In more mature, fit-for-purpose organizations (Maturity Level 3 and deeper) Upstream Kanban is defined and the commitment point is no longer ambiguous.
- The commonly recurring visualization patterns are three-board design, two-board design, and iterative discovery.
- The first three-board design was described and presented in Patrick Steyaert's Nemetschek SCIA case study (2009–2013).
- The Acceptance Criteria policy is an important element of a three-board design.
- The main issues that the three-board design addresses are: (1) lack of a proper discovery system that would support understanding the client's needs and (2) partial communication or lack of communication between discovery and delivery.
- The two-board design doesn't necessarily mean two separate boards. It just indicates that the Upstream and the Downstream parts are clearly defined and the commitment point is transparent and well-understood.
- The two-board solution is the consequence of an emerging Upstream process that results from scaling Kanban in width and from a maturing process.
- Upstream work may have an iterative nature. But even though the work in the iterations is repetitive, it's not identical. In each iteration, we collect a different set of data, which later informs the decision of what real option should be applied.
- When shared services are utilized by both Upstream and Downstream processes, it can cause problems with non-instantly available resources.

3

The Roles in Upstream Kanban

One of the Kanban Method's Change Management principles clearly states that you should "start with what you do now,"[22] which means:

- Understand current processes, as practiced.
- Respect existing roles, responsibilities, and job titles.

The famous "Blue Book," *Kanban: Successful Evolutionary Change for Your Technology Business* by David J Anderson,[23] made no mention of roles. The assumption was that employees would keep their current roles and job titles and, through acts of leadership, they would take on (sometimes additional) responsibilities related to managing flow. However, over the years of practicing Kanban and observing different implementations, it became clear that there was a need for a more prescriptive definition of roles. It was never the intention, though, to create new job positions or job titles! The roles were described to provide clear guidance on the set of responsibilities that needed to be covered to ensure the flow of work at different maturity levels (referred to as ML1 through ML6).

Kanban practitioners experienced multiple ways of implementing roles in their organizations, and those patterns are now codified in the *Kanban Maturity Model*, where you can read more about this.[24]

---

[22] https://all.kanban.plus/en/content/kanbanplus/bwk/default/bwk-posters/poster/kanban-in-numbers-2/9-principles

[23] David J Anderson, *Kanban: Successful Evolutionary Change for Your Technology Business* (Seattle: Blue Hole Press, 2010).

[24] D. J Anderson and T. Bozheva (2021), chapter 19, Barriers to Adoption.

This chapter focuses on two roles that support introducing and improving Upstream Kanban, the Flow Manager and the Service Request Manager.

Flow Manager

When organizations make a transition from Maturity Level 1, *Team-Focused*, to Maturity Level 2, *Customer-Driven*, we observe a shift from managing tasks to managing customer-valued and customer-requested deliverables. The focus switches to the flow of work rather than merely completing small tasks.

At ML2 we recognize that upstream and downstream flows may still be indistinguishable from one another and that the commitment point remains ambiguous. Or the Upstream system functions in a degenerate form of *To Do—In Progress—Ready for [downstream activity]* states in the process. Whatever the case, the Flow Manager is responsible for the end-to-end flow of work in its current shape with no distinction between Upstream and Downstream Kanban (Figure 23).

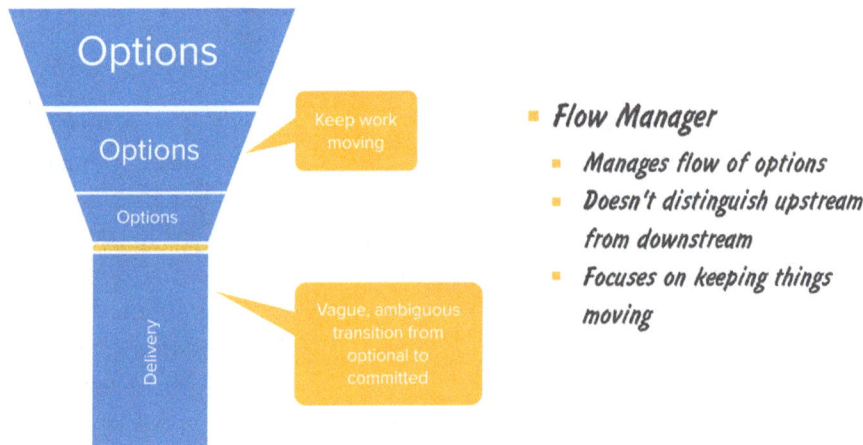

Figure 23 The role of a Flow Manager[25] in Upstream and Downstream processes

Implementation Guidance

- The Flow Manager is responsible for the following:
  - Create the consciousness that the service team is delivering a service to identified customers.
  - Ensure that flow metrics are collected (KMM practice MF 2.4).
  - Facilitate the workflow of the Kanban Meeting.
  - Facilitate understanding customer requests.

---

[25] D. J Anderson and T. Bozheva (2021), p. 182.
https://kanban.plus/blogs/blog/emerging-roles-in-kanban

- Facilitate the resolution of blockers, rework, and aging WIP–related issues that are escalated from the service team.
- A typical approach to implementing this role is to assign it to a team member who has volunteered for it and has the appropriate knowledge and skills to do the job.

Service Request Manager

For some number of years, the question of what to do with "middlemen" in the workflow has lingered. Generally, we wish to remove non–value adding middle positions (Figure 24) from the workflow. However, we also wish to avoid resistance to change. These are two core tenets of Kanban coaching as well as overall goals we might have for change management when deploying Kanban in an organization. The following guidance has existed since 2009: We seek to elevate the role of the middle person above the workflow, out of the value stream. The most common example of this is shown in the bottom portion of Figure 25.

- **Service Request Manager (SRM)**
  - *Manages flow of options*
  - *Upstream*

- **Service Delivery Manager (SDM)**
  - *Manages flow of committed work*
  - *Downstream*

Figure 24 The roles of a Service Request Manager[26] and a Service Delivery Manager in Upstream and Downstream processes

The Product Owner in the value stream:
- makes the prioritization decisions,
- is the single wringable neck,
- is inherently fragile,
- performs individual heroics, and
- is inherently part of a low-maturity organizational design.

---

[26] D. J Anderson and T. Bozheva (2021), p. 204.
https://kanban.plus/blogs/blog/emerging-roles-in-kanban

The product owner elevated above the value stream facilitates the replenishment meeting and risk management policies and:

- owns the risk-assessment framework,
- owns the triage-decision framework,
- facilitates decision making by others, and
- codifies decision making such that repeatable patterns of decision making are captured as policy. Codification facilitates scaling and makes the transition to newer personnel faster and more consistent.

The goal is to reposition the role of the Product Owner as a risk manager and facilitator: someone who owns the policies for the system, which puts decisions together

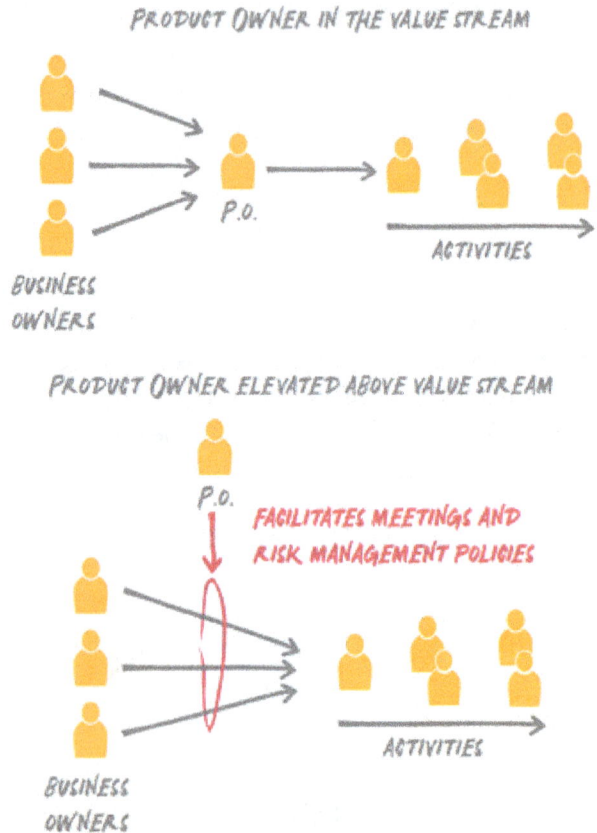

Figure 25 Product Owner in the value stream and elevated above it

with facilitating the decision-making mechanism. This role is of higher value, is transparent and open to scrutiny, and relieves us of the risk of the "hero Product Owner" who magically understands where the best business value is to be found. This elevated risk-management and policy-owning position improves corporate governance, improves the consistency of processes, and reduces the personal risk associated with a single individual making all the decisions.

When the Product Owner is successfully repositioned above the workflow as the owner of the policies for risk assessment, scheduling, sequencing, and selection, they have successfully transitioned into the Service Request Manager (SRM) role.

The role of Service Request Manager (SRM) emerges from the Flow Manager at Maturity Level 3, where the SRM acts as both the service delivery manager and risk manager for Upstream activities.

The name of the role—Service Request Manager—is flexible and can be adjusted to any name you see fit for your organization (e.g., Product Request Manager, Feature Request Manager, and so on).

Implementation Guidance

Typical functions of the SRM include the following:

- Develop an understanding of customers' needs and expectations.
  - Oversee the development of a consistent request elaboration process, agreed upon by all stakeholders. This includes defining explicit policies for triage, managing options Upstream, qualitatively evaluating options, and discarding Upstream requests.
  - Facilitate the Service Request Review (KMM practice FL 3.5).
  - Facilitate, select, and order work items at the Replenishment Meeting.
- At higher maturity levels, the SRM facilitates the Upstream Risk Review and participates in the Operations Review.
- The SRM ensures that the decisions made align with the organization's strategic objectives. Therefore, a solid understanding of the business, as well as well-developed communication and negotiation skills, are essential.
- The SRM role could be implemented as an individual or group responsibility, job title, or position in the organizational structure.

Takeaways

- One of the change management principles clearly states that everyone keeps their current roles and job titles. They take on (sometimes additional) responsibilities related to flow management through acts of leadership.
- Over the years of practicing Kanban and observing different implementations, it became clear that there was a need for a more prescriptive definition of roles.
- The roles were described to provide clear guidance on a set of responsibilities that need to be assigned to ensure the flow of work at different maturity levels.
- At Maturity Level 2, the Flow Manager is responsible for the end-to-end flow of work in its current shape, without distinguishing between Upstream and Downstream Kanban.
- The role of Service Request Manager emerges from the Flow Manager at Maturity Level 3, where the SRM acts as both the service delivery manager and risk manager for Upstream activities.

4

Upstream Feedback Mechanisms: Meetings and Reviews

This chapter provides a closer look at the Upstream meetings and reviews that are avenues for feedback and can serve as mechanisms for opening and closing feedback loops required for the Upstream system to operate effectively.

The goal of the *Implement Feedback Mechanisms* general practice is to enable both reflection against desired outcomes and the opportunity for adjustments, changes, and mutations to policies, processes, workflows, and working practices. Implementing a feedback mechanism is a key element to enabling evolutionary change.[27]

The feedback mechanisms are not limited to meetings and reviews. As Andreas Bartel says, "You do not have to have meetings to have feedback loops. I would even say that the feedback loops are agnostic to anything a human can invent or introduce in an organization because they exist anyway. Nonetheless, the real question here is not whether meetings are necessary or not. But rather how we can use meetings or different forms of communication for feedback loops, and what are the benefits, if we have any."[28]

How to Introduce Meetings and Reviews in Kanban

If you don't know whether introducing a new meeting to your calendar is a good idea, here is a short questionnaire you can use before doing so. Remember, your job is to find the lowest-resistance approach

---

[27] D. J Anderson and T. Bozheva (2021), p. 250.
[28] https://kanban.plus/pages/kanban-maturity-model and [28a] https://kanban.plus/blogs/blog/feedback-loops-everything-you-need-to-know-interview-with-andreas-bartel

to implementing each meeting given what you already have. Ask yourself the following questions about each feedback mechanism described in the KMM.

The First Set of Questions

- Do I have this feedback mechanism already? Should we recognize it for what it is and just tune it up a bit?

  It's quick and easy. If the feedback mechanism exists already (such as the daily status meeting), it already has its cadence. People are used to it, so you don't need to add any additional calendar-messing activities. Simply adjust your agenda and what you focus on in the discussion and the decision-making process.

- Do I have an alternative (that can be repurposed)?

  Do you have existing practices that could be adapted to provide opportunities for effective feedback loops? Importantly, review your existing meetings to evaluate their suitability. Do you practice Daily Stand-ups or Retrospectives? That's great! Don't rename them just to feel more "kanbanized." It's fine to keep an alternative; simply adjust its focus, agenda, and anticipated outcomes.

- Do any of them require improvement?

  To start, look at the Feedback Mechanisms' descriptions in the *Kanban Maturity Model* or on KMM+.[29] Do you think your existing meeting or review would benefit from some KMM-style improvements? Introduce them without changing anything else in the formal name or cadence of these appointments (if not absolutely required).

However, if you answered no to the first three questions, and your assessment says that you need a new feedback mechanism, ask yourself the following.

The Second Set of Questions

- Can I merge this feedback mechanism into any existing meeting or review?

  One of the most popular feedback mechanisms that trainees start practicing after completing the *Kanban System Design* class is Blocker Clustering. Very often, however, they don't experience so many blockers as to need a separate conversation about them; but on the other hand, they don't want the analysis to happen only every few months.

  A good practice in such a situation is, for instance, merging the blocker clustering and discussion about blockers into another existing mechanism—such as your improvements discussions or flow reviews.

If you answered no and see that you need to book a new time slot in the calendar, answer:

---

[29] https://kanban.plus/pages/kanban-maturity-model

- Can I merge two or more new feedback mechanisms?

  This one is very similar to the previous example, but here you merge a couple of new reviews into one. In Mauvius Group Europe, we conduct the Service Request Review and the Service Delivery Review at the same time due to the number of items in progress and the actors involved in both parts of the system.

  Merging several feedback mechanisms—several meetings and reviews—together is a common practice in small-scale implementations and smaller organizations. Implementing the full set of seven makes the most sense in organizations numbering several hundred or thousands of people.

However, if the answer is no again, then, and only then, create a new, separate meeting or review.

Don't Forget!

Feedback Mechanisms are not comprised only of what time employees of your organization spend in meeting rooms! Look around at the elements of your environment that "talk" to you, such as:

- Kanban boards
- Metrics
- Customer surveys
- Your colleagues sending instant messages

According to Andreas Bartel, "Feedback loops take place everywhere and anywhere, but to raise our self-awareness and have greater control over them we need to make them explicit. That is, have cadences." Table 1 (page 52) presents a summary of Upstream Kanban's formal feedback mechanisms.

Prerequisite Meetings and Reviews at Maturity Levels 1 and 2

As we discuss in Chapter 2, Designing Upstream Kanban, even those companies where the services are (in general) of lower maturity are already aware of the need for having a visual representation of the discovery process. Visualization alone is clearly not enough for the organization to mature, and it's critical to establish a proper feedback mechanism. At Maturity Levels 1 and 2 we still observe methods for sharing feedback that are supporting the organization's growth toward a full Kanban workflow.

Later, when the organization has already reached Maturity Level 3, those feedback mechanisms might still be a support for the nodes that remain separate units (e.g., even though there is a fully functioning Service Request Review or Service Delivery Review, the teams may still want to keep Team Retrospectives for their internal improvement purposes).

Prerequisite Meetings and Reviews at Maturity Level 1

Team Kanban Meeting

The Kanban Meeting[30] is intended to create a collaborative conversation about the status of the work, the impediments in the workflow, and problems in the emergent process—such as overloading some individuals—as well as to define appropriate actions for resolving the identified issues.

Whenever there is a specific team working in the Upstream part of the process (e.g., Business Analysts, Product Owners, Managers, Subject Matter Experts, etc.), the discussion about this process can be facilitated in the Team Kanban Meeting.

Similarly, whenever an Upstream-typical role (e.g., Business Analyst) is a part of the team, there is a good chance for at least a shallow version of the Upstream system in the team's workflow, and this work should be included in the Team Kanban Meeting process overview.

Implementation Guidance

- Hold the Kanban Meeting daily, always at the same time, to coordinate the work within the team and to facilitate self-organization.
  - Conduct the meeting in front of the kanban board. Make sure that all team members update the status of their work items and show them on the board before the meeting.
  - Walk the board from right (the part that is closest to completed work) to left (the part with not-yet-started work).
  - Report the status of each ticket and each work item on the board. Kanban Meetings always focus on the work, rather than the workload of individual people. Iterate across the tickets on the board, not around the group of people attending the meeting.
  - Keep the Kanban Meeting short by focusing the conversation on completing work items and resolving issues such as possible delays, blockages, technical problems, lack of information, and so on.
- Treat issues that require more time after the meeting, involving only the team members who can contribute to resolving them.

Team Replenishment Meeting

The purpose of the internal Team Replenishment Meeting[31] is to select work items from the backlog to commit next and to replenish the queue for delivery until the next Replenishment Meeting.

---

[30] D. J Anderson and T. Bozheva (2021), p. 252.
[31] D. J Anderson and T. Bozheva (2021), p. 254.

Conducting Team Replenishment Meetings supports establishing the first explicit policies of selection criteria. It lays the groundwork for a better future understanding of the work happening to the left of the immediate team's backlog.

Having a Team Replenishment Meeting and policies supporting decisions on "what we pull next" at the team level creates a solid shared understanding of pull decisions and a good base for further, more mature process development. Team members realize that the work cannot be pushed into the process and that before it is accepted, it should go through the (initial) Definition of Ready (read more in Chapter 2, Designing Upstream Kanban).

Implementation Guidance

Discuss topics related to the requested work:

- Information about work items ready to be pulled into the system
- Dependencies on other work items and technical risks associated with implementing the potentially selected work items
- Information needed to facilitate implementing the potentially selected work items
- Decisions about the order in which to pull the selected requests
- Select requests and then pull them into the *Next to Start* column of the kanban board.

Workflow Meetings and Reviews at Maturity Level 2

Workflow Kanban Meeting

The Workflow Kanban Meeting[32] is a higher-maturity version of the Team Kanban Meeting. Its scope includes the end-to-end workflow of a service or project. The intent of the practice is to involve the respective service teams in a collaborative conversation about the status of the work, queues between stages of the workflow, and problems that affect the flow of work. The identified issues are resolved after the meeting by team members with relevant knowledge and skills.

Implementation Guidance

Hold the Workflow Kanban Meeting with regular frequency. It's typically held in front of the board daily, at the same time, and should take about fifteen minutes or less.

- Conduct the meeting in front of the kanban board that visualizes the complete workflow. Make sure that the board has been updated and shows the real status of the work.
- Report the status of each ticket, each work item on the board. Kanban Meetings always focus on the work rather than the workload of individual people. Iterate across the tickets on the board, not around the group of people attending the meeting.

---

[32] D. J Anderson and T. Bozheva (2021), p. 256.

- Walk the board from right (the part that is closest to completed work) to left (the part with not-yet-started work).
- Discuss priorities and team member pull decisions. Are there any opportunities for collaboration? Treat issues that require more time after the meeting, involving only the team members who can contribute to resolving them.
- Discuss work items ready to be started. Do you recognize any dependencies or risks?
- Agree on temporary policy overrides when appropriate to manage risk.
- Keep the meeting short by focusing the conversation on completing work items and resolving issues such as possible delays, technical problems, lack of information, and so on.
- In some cases, it might be appropriate to combine the internal Workflow Replenishment Meeting and the Kanban Meeting.
- Service team members have access to the end-to-end workflow kanban board at any time. If they find issues in the workflow for which discussion cannot wait until the next Workflow Kanban Meeting, they take the initiative to signal and resolve them at once.

Supporting Case: Optimizely

Every Monday morning the team at Optimizely held what they called the planning session, which meets the definition of the Workflow Kanban Meeting (Figure 26):

- Designers, Researchers, Project Managers, Tech Leads, and occasionally the CEO took part in it.
- They went together through the system, checking the status of every item on the board.
- The new work was pulled.
- The meeting was a platform to discuss the work in progress and the importance of the work items.
- Cross-team dependencies were addressed and tackled.
- This meeting also served to better the understanding of who was doing what and what was most important.
- The key part of the meeting was the conversation and discussion of the workflow and work in progress, allowing everyone involved to come together and get clarity on the priorities.

Figure 26 Presenting the Workflow Kanban Meeting at Optimizely

Workflow Replenishment Meeting

The internal Workflow Replenishment Meeting[33] allows the service team (or the team of teams involved in the end-to-end service workflow) to select work requests and refill the *Next-to-Start* column of their kanban board so that they do not run out of work before the next Replenishment Meeting.

At ML2 we start observing the emerging Upstream system (even in its simplest form of one *Analysis in Progress* stage) and the initial forms of Definition of Done and Acceptance Criteria policies at the level of the whole workflow. This is what the mature Upstream system at ML3 will evolve from.

Implementation Guidance

Involve leads or representatives of the service teams who take part in the entire workflow. If possible, involve the customer in the Workflow Replenishment Meeting.

- Select requests solicited by the customer and place them in the *Requested Work* column.
- Make sure that the service team understands the requested work.
- Consider technical aspects and dependencies among service teams when selecting a request to be pulled into the *Next-to-Start* column.

---

[33] D. J Anderson and T. Bozheva (2021), p. 255.

Supporting Cases

More frequent replenishment enables making appropriate decisions based on the current business situation. Therefore, more frequent replenishment is more agile. On-demand replenishment is most agile.

However, we always need to consider two other factors: the arrival rate of new information and the cost of holding a meeting.

Scenario 1: On-Demand Replenishment

Working in investment bank, I was responsible for managing the project while also playing the role of a Product Owner (with authority limited to deciding on "what we do next"). With a very strict deployment calendar known in advance for the whole calendar year, we were able to anticipate the arrival pattern of a specific demand. Normally, in the first three to five days after releasing new products, we would receive some failure demand coming back in the form of small change requests or defects. While trying to solve the root cause, we also dealt with them regularly to avoid them piling up and creating technical debt.

The project team worked most of the time on the more time-consuming work items (with seventy-five days lead time at 85 percent), and the decision on what to do next was discussed and approved by business partners in the official communication. The after-release defects and change requests, however, had a very high turnover rate, and we were authorized to decide on them by ourselves. Solving those work items usually took between twenty minutes and three hours, and by having a few people focused on them, we were able to fix everything in one to two days.

However, the work-in-progress limit (five) in the *Next* column was established to cover items taking much more time, and in post-release periods we frequently ran out of pullable work very quickly—sometimes a few times a day. Of course, we could temporarily change the WIP limit, but keeping it at the same level was much easier, and it increased focus on those five tickets to pull next. Over time, project members developed the habit of picking up the phone whenever two items were remaining in the *Next* column: Sitting at my desk on the third floor, I heard, "Anna, what do we do next?" on the other end of the phone, and I knew it was time to take a short trip one floor down. We held a quick, five-minute on-demand Replenishment Meeting, and everyone knew what to do next.

We could do this because:

- The coordination cost was close to zero (a few minutes of people's attention and some calories burned while running up and down stairs).
- We had full authority to decide what to do.
- The arrival rate of new information was very high.

Scenario 2: JP Morgan Chase

Working with JPMC, Abbigayle Wall discovered distortion in the flow of work caused by improper planning and a mix of work types in the work cycle. Whenever there was too much focus on a specific work item type in an upcoming cycle, the lead time of the other work suffered (e.g., focus on technical work slowed down project work and vice versa).

Because of this, they introduced a pre-planning meeting (very short, about fifteen minutes) a day before the official meeting, where information about available kanbans was passed to the team. It was supported by an explicit work-allocation policy. The work during the Replenishment Meeting was pulled in a round-robin format, based on the pre-established and previously announced number of available slots. This way, team members could stay focused during the planning (Replenishment) meeting and run it effectively.

Flow Review

The intent of the Flow Review[34] (FR) is to develop an initial understanding of the delivered service and use it to facilitate work planning and thereby improve predictability.

At Maturity Level 3, we observe the Service Delivery Review (FL 3.4) and Service Request Review emerging from the Flow Review. The difference is that an FR is inward facing: It looks at capability without considering customer needs or expectations. With an FR, more is usually better, and the intent is to become as effective as possible.

Flow Reviews develop a quantitative understanding of the current capability of the service team. If a target is defined and known, it might be compared with the system's capability to trigger improvement actions. At this level of organizational maturity, the service team is not so concerned with where the target came from—it may simply be a goal set by a manager to encourage improvement. Tying objectives to real customer expectations and business risks emerges at Maturity Level 3.

A Flow Review is typically held twice a month, facilitated by the Flow Manager or their immediate superior. Other participants are typically the workers from the service delivery workflow who can contribute to analyzing and understanding certain events in the workflow, such as long-time blockages or work items spending a long time in progress. There is little or no external representation at this meeting.

Implementation Guidance

Before the meeting, review the progress and system capability data derived from the kanban board or tracking system, such as the following:

- Distribution of demand per work type:
  - Average arrival rate of the demand: the average number of requests for a time period per work type
  - Average delivery rate (throughput): the average number of completed work items for a time period per type of work, using the same unit of time for both the arrival and delivery rates

---

[34] D. J Anderson and T. Bozheva (2021), p. 256.

- Average lead time per type of work: the amount of time between selecting an item to work on and delivering it to the customer
- Average customer lead time per type of work: the time measured from the moment the customer placed the request to the moment it was delivered
- Range of blocker time per blocker cause
- Range of defect and rework time per cause
- Highly aged work items
- During the meeting, reflect on the collected data and analyze the cases for which there was a perception of poor service or a failure to meet customer expectations, if known.
- Define actions to improve the flow. Start with defining actions for reducing blocker time, rework, or WIP aging.
- Provide feedback and communicate decisions to the service team and their immediate manager.

Upstream Meetings and Reviews at Maturity Level 3

Replenishment Meeting

The purpose of the Replenishment Meeting[35] is to decide, together with the customer and business stakeholders, which work requests to select for the next period until the next Replenishment Meeting. The selection of work items considers the information about the service capability and the available capacity; this is brought to the meeting by representatives of the service team.

The typical participants in a Replenishment Meeting are the customer, decision makers, and other relevant stakeholders.

The Replenishment Meeting should have an established cadence. Two factors must be considered when determining the frequency of the meeting: the arrival rate of new requests and the cost of holding the meeting. More frequent meetings with customers and business stakeholders enable us to make appropriate decisions based on the current business situation. Therefore, more frequent replenishment is more agile. On-demand replenishment is most agile.

Implementation Guidance
- Prepare the Replenishment Meeting:
  - Identify what requests are available and need to be discussed at the meeting. Collect and visualize these requests in a queue to be reviewed during the meeting.
  - Prepare data and information needed for the meeting, such as service-related data.

---

[35] D. J Anderson and T. Bozheva (2021), p. 260.

- Conduct the Replenishment Meeting:
  - Discuss what requests and combinations of requests can be selected for delivery before the next Replenishment Meeting.
  - Select requests and pull them into the *Selected for Delivery* column of the kanban system. Make sure that the selected items meet the acceptance criteria. Document what criteria or additional considerations for selecting requests are being used if these differ from the established policies.
- Bring Replenishment Meeting–related observations to the Service Delivery Review for further analysis of the system.
- The essential aspect of the Replenishment Meeting: It's the point at which the customer commits to the outcomes of the requested work and the service team commits to delivering it within the customer's expectations. Before this point, the customer's requests are considered options that can still be discarded.

Service Request Review

The intent of the Service Request Review[36] is to understand the status of the requests in the Upstream Kanban system and select those that need to be moved across the commitment point and those to be elaborated (analyzed or synthesized) further.

Typically, participants in the Service Request Review are the customers or their representatives. The frequency of the review can be biweekly or monthly and must be established based on the transaction and coordination costs associated with conducting the review.

Implementation Guidance

- Prepare the review:
  - Review the status of the work requests that have been elaborated.
- Conduct the meeting:
  - Walk the discovery board from right to left.
  - Items to the right may be ready to commit. Items further to the left may be ready to be promoted to the next level.
  - Identify the requests that are ready to commit. Consider the following aspects when making the selections:
    - Request acceptance criteria
    - Proper diversity of options for each work type or customer
    - Dependencies among the requests
    - Important dates and delivery times

---

[36] D. J Anderson and T. Bozheva (2021), p. 262.

- ◇ Available capacity of the delivery kanban system as reported by the Service Delivery Manager
- ◇ Cost of delay and required class of service of the item, and how this affects dependencies
- Identify items that may be ready to be promoted to the next stage, should be shelved until circumstances change or new information arrives, or should be discarded.
- Make sure that enough work requests are selected to guarantee that the delivery kanban system does not starve. Likewise, each Upstream stage needs enough options to ensure choice Downstream. There should always be a certain discard rate.
- Review the discard rate. Is it in line with expectations? High uncertainty would mean a higher discard rate, perhaps as high as 90 percent, while low uncertainty would mean a low discard rate, perhaps as low as 1 percent. If the discard rate is out of line with expectations, there are two possible problems: insufficiently good ideas or ineffective means to evaluate good versus bad options to be promoted to the next stage. Refer the results to a Risk Review.

Service Risk Review

The intent of the Service Risk Review[37] is to understand and respond to risks to the effective delivery of products and services.

This review is typically held monthly and is facilitated by a Service Delivery Manager, Service Request Manager, Director, or a Kanban coach. Other participants include anyone with information or experience of recent blockers, Project and Program Managers, customer-facing managers, and managers from dependent services. The scope of the review is one or more kanban systems being used for service delivery.

Implementation Guidance

- Prepare the review:
  - Review the status of the issues and risks from previous Service Delivery Reviews, Service Request Reviews, and Delivery Planning Meetings.
  - Prepare the list of potential problems to be discussed at the review.
  - Risk review can consider both Downstream delivery and Upstream discovery workflows.
- Conduct the review:
  - Categorize the identified risks.

---

[37] D. J Anderson and T. Bozheva (2021), p. 263.

- Analyze the likelihood and impact of the risks for meeting customer expectations identified in the Delivery Planning Meetings, Service Delivery Review, and Service Request Review.
- Analyze other relevant aspects of the identified risks; for example, those that address nonfunctional or safety expectations of the customers.
- Define appropriate actions to mitigate the identified risks based on the outcome of analyzing them.
- After the review, communicate decisions as appropriate and make sure that affected artifacts (kanban system designs, policies, etc.) are properly updated.

Table 1 Summary of Feedback Mechanisms

Feedback Mechanism	Objective and Benefits	Input	Output	
Prerequisite Meetings at Maturity Level 1				
Team Kanban Meeting	Create a collaborative conversation about the status of the work, the impediments in the workflow, and problems in the emergent process.	The status of the work items is updated by all team members and visible on the board before the meeting.	Resolving issues such as possible delays, blockages, technical problems, lack of information, and so on.	
Team Replenishment Meeting	Select work items from the backlog to commit next and replenish the queue for delivery until the next Replenishment Meeting.	Work items ready to be pulled into the system and information about them	Requests selected and then pulled into the *Next to Start* column of the kanban board.	
Workflow Meetings and Reviews at Maturity Level 2				
Workflow Kanban Meeting	• Involve the teams in a collaborative conversation about the status of the work, queues between stages of the options workflow, and problems that affect the flow of work. • Supports cross-team collaboration.	• Commitments • Decisions about new initiatives and capacity allocation • Policy changes • Updated board showing the real status of the work	• A shared understanding of how best to flow current work • Improvement suggestions	
Workflow Replenishment Meeting	Allow the service team(s) to select work requests and refill the *Next to Start* column so that they do not run out of work for the next period.	• Observations resulting in changes to behavior or policies at Replenishment Meetings • Policy changes/portfolio changes	• Decisions regarding what to pull • System changes (to Kanban Meeting)	
Flow Review	• Develop an initial understanding of the delivered service and use it to facilitate work planning and thereby improve predictability. • Develop a quantitative understanding of the current capability of the service team.	Progress and system capability data derived from the kanban board or tracking system	• Defined actions to improve the flow (e.g., reducing blocker time, rework, or WIP aging) • Feedback and decisions to the service team and their immediate manager	
Upstream Meetings and Reviews at Maturity Level 3				
Replenishment Meeting	Decide, together with the customer and business stakeholders, which work requests will be selected for the next period.	• Available requests that meet selection criteria • Prepared data and information such as service-related data	• Customer commitment that they want the requested items; team commitment to deliver within expectations • Observations to the SDR for further system analysis	
Service Request Review	Understand the status of the requests on the Upstream kanban board and select those that need to be moved across the commitment point and those to be elaborated further.	Review the status of the work requests that have been elaborated.	• Enough work requests selected to guarantee that the delivery kanban system does not starve • Observations and policy changes for Service Risk Review	
Service Risk Review	• Understand and respond to risks to effective delivery of products and services. • Risk review can consider both Downstream delivery and Upstream discovery workflows.	• Review the status of the issues and risks from previous Service Delivery Reviews, Service Request Reviews, and Delivery Planning Meetings. • The list of potential problems to be discussed at the review.	Communicate decisions as appropriate and make sure that affected artifacts (kanban system designs, policies, etc.) are properly updated.	

Who should attend?	During the meeting/review	Frequency	Duration
Prerequisite Meetings at Maturity Level 1			
The Team Lead, Process Owner, team members	• Conduct the meeting in front of the kanban board that visualizes the complete workflow. • Focus on work rather than workers. • Walk the board from right to left. • Report the status of each ticket and each work item on the board. Iterate across the tickets on the board.	Daily	10–15 min.
The Team Lead or the person who is in contact with the customer, team members	• Discuss information needed to facilitate implementing the potentially selected work items. • Decide the order in which to pull the selected requests.	Weekly or as needed, based on the arrival rate of new information	20–30 min.
Workflow Meetings and Reviews at Maturity Level 2			
Flow Manager, immediate group or team doing the work (4–50 people)	• Conduct the meeting in front of the kanban board that visualizes the complete workflow. • Focus on work rather than workers. • Walk the board from right to left. • Report the status of each ticket and each work item on the board. Iterate across the tickets on the board.	Depending on the coordination cost; start with weekly	10–20 min.
Leads or representatives of the teams involved in the workflow; if possible, involve the customer.	• Select requests solicited by the customer and place in the *Requested Work* column. • Make sure the requested work is clearly defined. • Consider technical aspects and dependencies between service teams.	Start with weekly, but consider the arrival rate of new information.	20–30 min.
• Flow Managers or their immediate superiors and the workers from the service delivery workflow who can contribute to analyzing and understanding certain events in the workflow • Little or no external representation at this meeting	Reflect on the collected data and analyze the cases for which there was a perception of poor service or a failure to meet customer expectations, if known.	Typically, twice a month	30 min.
Upstream Meetings and Reviews at Maturity Level 3			
SRM, SDM, customer and business stakeholders, and service delivery representatives	• Discuss what (combination of) requests can be selected for delivery. • Pull selected items. • Document criteria used and additional considerations.	Once to twice a month	30 min.–1 hr.
SRM, customers, and their representatives	• Walk the discovery board from right to left. • Identify the requests that are ready to commit. • Identify items that may be ready to invest, should be shelved, or should be discarded. • Review the discard rate.	Once/twice a month	30 min.–1 hr.
SDM, SRM, director, or **Kanban coach,** anyone with information or experience of recent blockers, project and program managers, customer-facing managers, and managers from dependent services.	• Categorize the identified risks. • Analyze the likelihood and impact of the risks for meeting customer expectations. • Analyze other relevant aspects of the identified risks. • Define appropriate actions to mitigate the identified risks.	Monthly	1 hr.

Takeaways

- The goal of the *Implement Feedback Mechanisms* general practice is to enable reflection against desired outcomes and enable the opportunity for adjustments.
- A feedback mechanism is not limited to meetings and reviews.
- Don't rush with introducing a new meeting or review when you can introduce an alternative solution.
- Prerequisite meetings and reviews are the Team Kanban Meeting and Team Replenishment Meeting.
  - At Maturity Level 2, those are obviated by the Workflow Kanban Meeting, Workflow Replenishment Meeting, and Flow Review.
- Upstream Meetings and Reviews at Maturity Level 3 are the Replenishment Meeting, Service Request Review, and Service Risk Review.

5

Upstream Metrics

Another feedback mechanism that we can use is metrics. In this chapter, we take a closer look at different types of metrics that you can apply with Upstream Kanban to effectively support the decision-making process.

Quality-Related Metrics

Failure demand represents demand generated from previous poor-quality deliverables or demand that never should have been received. Failure demand is a result of low quality and is avoidable if the initial quality is better matched to customer expectations (see the box entitled The Effect of Failure Demand on Scaling on page 57).

Some examples of failure demand include the following:
- Defect-fixing requests
- Rework due to usability problems
- Rework due to poor design or not understanding customer needs
- Features requested by users because other functionality did not work (workarounds)
- Demand that never should have been accepted (It is failure demand because the failure was in accepting it. It is a failure of product management, strategy, triage, etc.)

Disruptive demand represents the demand that disturbs the flow of work but does not bring any additional value when fulfilled. Some examples of disruptive demand are:
- Non–value adding estimation requests: requests for precise capacity allocation per man-days or man-hours available for the

project at the beginning of the process, very often right after the ideas are dropped in the pool of ideas or backlog.

- Expediting items just to speed them up—with no regard for policies—that results in wasted time and context switching

Speculative demand represents demand for which:

- The true value is unknown or unknowable.
- The request is always urgent and important.
- The requirements are very poorly defined.
- There is a high probability that the work will be abandoned or aborted (but there is serious denial that this happens, or there is no tracking or reporting, so no visibility of this problem).

Speculative demand tends to come from Upstream and disrupts the Downstream system. It is important to track these as first-class work items and have a type for them (e.g., a *Request for Estimation* is something that should be tracked so that we have visibility on it so that we can assess the impact of this type of demand).

Even though speculative and disruptive demand may seem to overlap with some of the characteristics of failure demand, it is worth remembering that speculative demand is an original demand, whilst failure demand is a derivative demand.

Speculative demand, however, can be one of the root causes of failure demand.

[ML3] Measuring Quality-Related Metrics

Quality of service: The rate of requests delivered without needing rework.

- Measure the number of work items delivered without needing rework (done right the first time).
- Divide this number by the number of all work items and multiply by 100 percent.
- If you have identified work item types already, remember to measure quality per work item type.

Failure demand/value-adding demand rate:

- Measure the number of work items that you labeled as failure demand.
- Measure the number of work items delivered without needing rework (done right the first time).
- Divide failure demand by value-adding demand and multiply by 100%.

Failure, disruptive, and speculative demand per the root cause:

- Collect information about failure/speculative/disruptive demand root causes.
- Cluster them and bring the results of the analysis to the Service Delivery Review.

The Effect of Failure Demand on Scaling

At Maturity Level 3, product design, quality, and service delivery are all within customer expectations and tolerance levels. The organization is considered trustworthy. The customers are satisfied. The customers' expectations are met regularly and sufficiently well such that customers are satisfied with the level of service. Both processes and outcomes are consistent. The organization has developed the capability to respond quickly to changing customer expectations. Hence, failure demand drops dramatically when you achieve Maturity Level 3.

Problems with Low Quality

Failure demand is a result of low quality. If you spend a lot of time fixing things, reworking issues, and dealing with complaints from unhappy customers, then you cannot scale. When you are constantly in firefighting mode, fixing problems, you cannot scale. What you should do and focus on is finding new clients or new products to create and sell, adding value. However, when you are continuingly dealing with previous customers who complain, it affects your ability to sell new products. Low quality will take you a step back on many fronts, especially when it comes to referrals. When someone asks, "Can I speak to a previous customer?" you probably will not be willing to share their contact details.

Scale by Fixing Quality Issues

You scale by fixing quality issues and delivering products or services that meet customer satisfaction, not by dealing with complaints or non–value adding work items. You scale by freeing your processes of all the noise and distraction.

Having a kanban board helps to minimize the time spent on "managing" things, like walking around and asking for the work status. It gives you the opportunity and time to look left to see what is coming. You don't need to fight fires or micromanage workers. The board gives you time and intellectual capacity to anticipate.

Maturity Levels 0–2 are reactionary. Maturity Level 3 is a transition toward an anticipatory organization that we observe beginning at Maturity Level 4. You can't be anticipatory until your process is free of all the immediate distractions. You can achieve a greater level of scale by removing or fixing all the quality issues.

If you do things with high quality, your customers trust you. The level of trust goes up with organizational maturity. When you can trust one another within the organization, you don't need to supervise every single step, and you can focus on value-adding work. Hence, we scale by improving organizational maturity.

Implementation Guidance

- Explicitly visualize work items that represent failure demand.
- Collect the tickets that represent failure demand (or register them in whatever tool you use) for deep analysis of the root causes and the parts of the processes that produce it.
- Periodically analyze and quantify the amount of failure demand and its impact.
- Report the analysis results during the Service Delivery Review.
- Through a review at an appropriate level, such as Service Delivery, Risk, or Operations Review, consider revising policies to reduce the failure demand.

[ML2 and ML3] Abandoned and Discarded Work

Abandoned Work

At Maturity Level 2, where the commitment point is still ambiguous and the process is not clearly divided into Upstream and Downstream, we measure the abandoned work. What we typically see is work that has gone stale and is effectively abandoned, but this has not been recognized as such.

By abandoned work, we refer to work items that:

- work had started on, and the lead time had started being calculated

  or

- were rejected from the process before completion of the work.

Abandoned work (at ML2 or below) is work that the customer believes has been committed (or started), but as yet nothing has happened on it. It may have been actively abandoned (closed), but more likely it is simply open and stale with no work occurring on it. One way to deal with abandoned work is to set a "guillotine" (a cutoff date) whereby "if this item has not been updated since [number of days/weeks/months], we consider it abandoned and close it as such."

Abandoning work is particularly a problem if the Upstream generates disruptive, speculative demand Downstream, sucking away valuable capacity. It is desirable to avoid using Downstream capacity (whenever possible) for work that is speculative with a high possibility of discard. As defined, the bottleneck or capacity-constrained resource should always be Downstream. Upstream must always have the slack capacity (time spent on options that will be discarded), because disruptive, speculative demand sucks away precious capacity-constrained Downstream resources.

The time spent on processing items that were started but abandoned before the work was completed is a waste. What is more, there is a lost opportunity to deliver work that could have been completed and would deliver value to the customer. Hence, abandoned items should be tracked and measured to better understand the root causes for rejecting the work and to improve the policies.

Implementation Guidance

[ML2] Measure the ratio of abandoned work items versus all items.

- Measure the number of work items that were rejected.
- Divide this number by the number of all work items and multiply by 100 percent.
- Collect information about root causes for abandoning the work and present the ratio per root cause.
- Bring the results of the analysis to the Flow Review.

Discarded Work

[ML3] Frequently, organizations have many submitted ideas that linger without attention for months—or even years. There is never enough capacity to start working on all the ideas and to get them done. Plenty are discarded sometime later because they no longer interest customers or fit the organization's business. However, the elaboration, development, and management of these ideas have already consumed considerable time and effort.

Acknowledging this fact shows significant progress. It is useful to define explicit policies for actively closing aged options before the kanban system's commitment point (i.e., Upstream) (Figure 27). This allows the organization to focus on ideas that are expected to bring more value to customers and to avoid discarding work items in later stages of their development, which wastes time and resources.

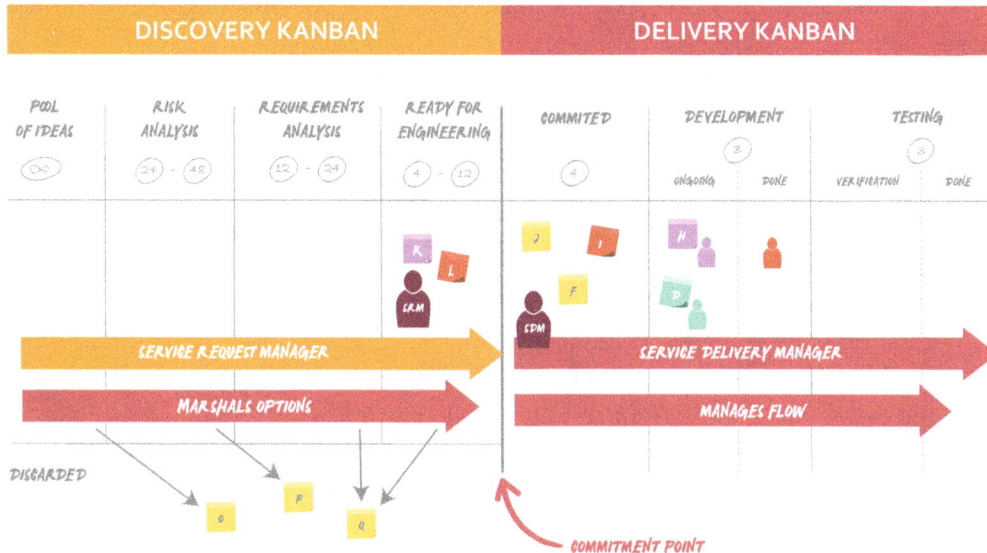

Figure 27 An example visualization of the placement of discarded items on the kanban board

Implementation Guidance

- Create a space on an Upstream/discovery kanban board or a bin to the left of the commitment point on a Delivery kanban board and actively display recently discarded tickets.

- Before removing the discarded options' cards from the board, review and reflect on what you have learned from the experience:
  - Could any of them have been a better choice?
  - Should this item be discarded?

- **Measure the discard rate:** the number of discarded tickets versus the total number of work items.
  - Establish a healthy range for the discard rate in your business context.
  - Regularly check whether the discard rate remains in the healthy range. Address any deviations from the range.

Lead Time at Lower Maturity Levels (ML1–2) as a Prerequisite

We define **lead time** as starting at the mutually agreed commitment point and continuing until an item is ready for delivery. This definition is unambiguous and works consistently for any workflow or service.

[ML1] **Team cycle time** is the time that passes from the point when the team pulls the ticket and starts working on it (at the Team Replenishment Meeting) until the ticket is ready for delivery.

[ML2] **System lead time** (health indicator or improvement driver) is the time that passes from the point when the delivery team agrees to pull the ticket and start working on it (commitment point) until the ticket is ready for delivery (i.e., has reached the delivery point). The delivery team is not responsible for any delays incurred by the customer when an item is ready for delivery. A few teams may be involved in delivering the work from the commitment point to the *Ready for Delivery* stage; hence, the system lead time may consist of a few different team lead times.

At lower maturity levels, the commitment point is often ambiguous and at best asynchronous; that is, the customer or the Upstream part of the workflow commits a request before the delivery or Downstream service workflow is ready to commit. In this type of asynchronous commitment, we measure lead time from the second half of the commitment, the point when the delivery workflow pulls the item into WIP. In Maturity Level 2 workflows, often there isn't a strong concept of commitment, so there is a tendency to measure lead time from the point a work item is submitted. This is legitimate for internal shared services and other similar systems where the work is irrefutable.

The difference between **customer lead time [ML2]** at lower maturity levels and system lead time is always a non-deterministic period of waiting on the front end, the time between when the customer submits a request—or believes it was committed—to when it is pulled into the kanban system and actually is committed. The effect of a difference in customer versus system lead time is that the customer lead time suffers from a longer, fatter tail. For a detailed explanation of lead time, including fatter versus thinner tails, see the Appendix.

Lead Time and Option Expiry Date at ML3

[ML3] **Lead time** (fitness criterion) is the time that passes from the point when the customer believes their order has been accepted (i.e., it has reached the request point) until the item is ready for delivery. In other words, we start counting lead time from the point when a customer can legitimately expect us to work on an item until we can legitimately expect the customer to take delivery. Any additional waiting time on the delivery end is not counted. In a Maturity Level 3 or deeper Kanban implementation with synchronous commitment, customer lead time equals system lead time: They are equivalent. Hence, there is no differentiation between lead times from Maturity Level 3 onward.

Should We Measure Lead Time in Upstream?

The work in the Upstream is optional. To have options is to have choices, which means that any item in the Upstream system may be subject to discarding. So, should we even bother measuring the lead time of a work item that ultimately can be removed from the system in the process of Upstream decision making? It depends on the domain we are in. So, let's clarify those dependencies.

Before we decide whether or not we should take care of measuring the lead time in the Upstream system, we should first understand the nature of our work and the domain we are in (Extremistan versus Mediocristan).

Mediocristan or Extremistan?

Mediocristan is a boring world of normal, predictable, and repetitive events. Within an understood range of variability, everything is expected; there are no surprises; and the impact of whatever may happen is very small. Huge changes with high impact don't exist in Mediocristan.

When describing how tall people are, what their weight is, and what the temperature in Central Europe during summertime is, you would navigate in a Mediocristan world. While it is boring to live in, Mediocristan is paradise. And it is the ultimate goal for having an answer for all the people asking you, "So, when it will be done?"

In a Mediocristan environment, we can easily and safely use averages in our attempts to anticipate future events, as this is where the Gaussian distribution is applicable, as shown in Figure 28.

Right Skewed	**Gaussian**	**Left Skewed**
(Weibull 2.0 < k < 4.0)	(Weibull k = 4.0)	(Weibull 2.0 < k < 4.0)

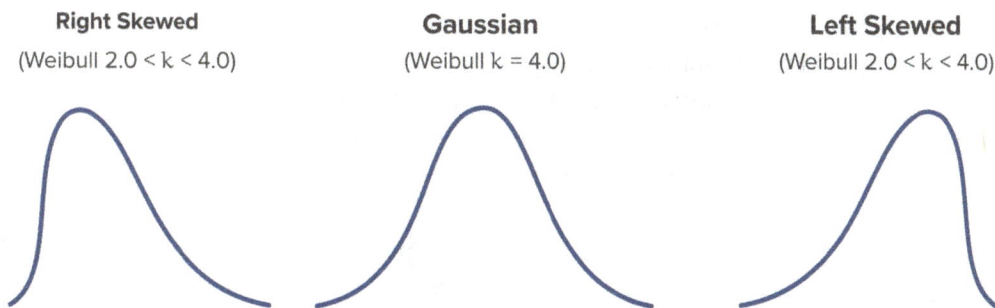

Figure 28 Gaussian distribution is applicable in a Mediocristan environment.

The range is very limited. The tallest person is less than double the average, and the shortest is a third of the average. The vast majority of people are very close to the average, with very few near the extremes. It would be impossible to find a person a thousand times taller than another.[38]

---

[38] https://all.kanban.plus/en/content/kanbanplus/kmm/default/kmm-posters/poster/kmm-metrics-poster/metrics-5-eng

Mediocristan means selling coffee in a city center cafe on a weekday, or ice cream on a beach in the summer. Demand will vary a little, but it is bound to a minimum and maximum expectation, and the variation tends to be distributed in a bell curve. In statistics, this is known as Gaussian normal distribution and is often represented by a bell-shaped curve.

In real life, we know that kanban system delivery rates (also known as productivity rate, velocity, or throughput rate) are Gaussian distributed. We also know that the liquidity of a kanban system—the volatility in the rate of pull within the system—is Gaussian distributed.

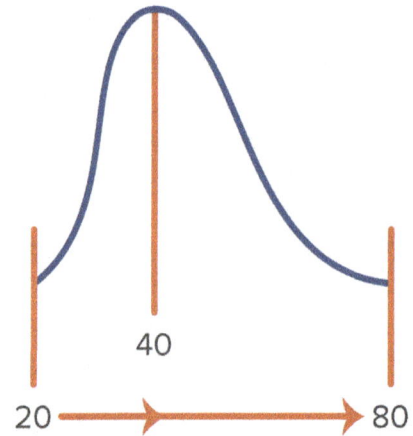

Figure 29 Typical Gaussian functions observed in service delivery and project management

Typical Gaussian functions observed in service delivery and project management look like the one in Figure 29, right skewed and exhibiting a spread of two times above and below the mean.[39]

Extremistan is a different world. This is where the unpredictable happens. Those rare and unpredictable events have a very high impact on the average. If you are at an event for self-published book authors, the average number of sold books per author will be 250, with an average annual revenue of $1,000.[40] But you have invited J. K. Rowling to give a keynote speech. By 2018, she sold 800 million *Harry Potter* books worldwide. This single number heavily affects the average, and now all of you in the room are millionaires (on average . . .). You can observe a similar situation in the movie industry and in the computer games and general entertainment markets.

Extremistan is publishing children's books and not knowing whether your new manuscript is the next *Harry Potter*. There is no concept of upper and lower limits; variation can be from zero to almost infinite and is not predictable. Extremistan is the world of Pareto distributions.

Pareto-class distributions are often referred to as power laws. These are very fat-tailed and represent a high risk. Hence the nickname Extremistan. In this range, tails exceed one hundred times the mode.

Fortunately, Pareto distributions rarely, if ever, appear in project management or service delivery problems. Pareto distributions do exist in product management problems such as payoff uncertainty; for example, how much will our next book make? Is it the next *Harry*

---

[39] D. J Anderson and T. Bozheva (2021), p. 393.
[40] https://www.alphapublisher.com/post/how-much-does-the-average-self-publisher-make; https://all.kanban.plus/en/content/kanbanplus/kmm/default/kmm-posters/poster/kmm-metrics-poster/metrics-5-eng

Potter? Or will it sell fewer than 300 copies? Pareto distribution problems are also known as *black swan* problems.[41]

Lead Time or Option Expiry?

Understanding which domain you are in will help you with deciding whether or not you should measure the lead time in the Upstream. Or, rather, what should you take care of?

If you are in a Gaussian domain (Mediocristan), you are interested in *concept-to-cash*. There is a conceptual idea in our *Pool of Work*, and this idea needs to go through the whole decision process until we decide that we either invest in it or discard it. This decision process involves activities like analyzing risk, preparing specifications and requirements, drafting the first design or proof of concept (PoC), and product marketing work. Then, eventually, the idea hits the commitment point when we decide that this is what we want and need to build. There is still an optional period (Upstream) and there is a committed delivery period (Downstream), but this whole (lead) time matters.

If we are designing and introducing new models of cars, we can expect that our Upstream (market research, ideation, concept, options exploration, simulation and proto-typing, customer focus groups, and so forth) will take a period of time that is bound with a minimum and a maximum. We may have an expectation that an existing model is replaced in our product range within three-and-a-half to seven years. History will have shown us that we can bring new models to market from conceptualization to realization within this time window. We know that a market exists for the existing model, and we can anticipate a certain sales volume for the replacement model within some lower and upper bounds. The range is always bounded by the size and population of the market and the specific size of the segment the model may be targeting.

We have an expectation of how much information we can learn from each stage and the cost (and time) to collect that information. Hence, our Upstream, while the work is optional, still expects lead time, and the total concept to cash is something we can measure.

In this (Gaussian) type of domain, you develop products because there is a gap in the market that you want to fill—if you can define a market based on a problem. If you can define a problem, can you measure (or at least model) the pain it is causing? And can you do this over time? If so, you can predict the cost of delay within some range of probability. If it takes you five years to solve the problem or deliver a product, it is five years of delay for which you can define the total cost and the rate of accruing the cost over that time and how that rate might change during that time. This is when the lead time in the Upstream matters. If you have some certainty that there is a problem that needs solving and that there is an incremental move from where you are now, and if you know how to reach that specific market where the gap occurs, then you care about how long that problem's solution is in the Upstream. You are dealing with something that you know has a cost of delay.

---

[41] https://en.wikipedia.org/wiki/Black_swan_theory; D. J Anderson and T. Bozheva (2021), p. 400.

If, however, you are in a Pareto payoff domain (Extremistan), you care about an option expiry and an investment: How long will you spend investing in it? How much money are you going to spend on it? You can't do either infinitely, as you don't have infinite time or money.

This is when you are not measuring the lead time, but instead you consider the option expiry (the point at which it is no longer viable to make a positive choice) and the number of options. How do you know whether your new manuscript is the next *Harry Potter*? You don't! What you need are lots of options. If, as a publisher and book editor, you knew that the manuscript you were holding would become the next *Harry Potter*, you would expedite it to market! The reality is that you do not and cannot know in advance whether a manuscript will turn into the next best-seller; the next blockbuster movie; the next craze, fad, or viral phenomenon. So, in such markets, cost of delay is incalculable and useless as a measure, metric, or decision-making tool.

Apply Filter: Cost of Delay

There is a difference between "Is this manuscript a new *Harry Potter*?" and "Can you launch a new application?" The first question has no cost of delay; it could be a total flop, and you sell 250 copies. If you knew it was another *Harry Potter*, there would be a cost of delay. However, you can't know that. It's an unknowable fact. The second question might have a clearer answer.

Ask yourself a simple filtering question: Can we identify and describe the cost of delay for this idea? The number can be within a range, or it might be a qualitative description. The most important thing is that you know what the cost of delay is.

If the answer is yes, you care about the Upstream lead time.

If the answer is no, you don't care about the Upstream lead time; you care about option expiry and the number of options.

The idea of Gaussian versus Pareto distribution and the idea of asking whether or not you can identify the cost of delay are very strongly correlated. If something has a Pareto payoff, you don't know what it is worth, so you can't calculate the cost of delay. If you can calculate the cost of delay, most probably the payoff is Gaussian.

The time items spend in the Upstream, in the *Ready* column, plus the delivery time, give you the concept-to-cash—when you can identify the cost of delay—and this is when you can measure the whole lead time. When you care about concept-to-cash, probably you will create a time guillotine on the *Ready* column, because you want the time items spend in this column to be short. By guillotining it, you keep the tail thin on the concept-to-cash distribution. If you don't guillotine it, you've potentially fat-tailed your concept-to-cash histogram.

If you don't know the cost of delay, you don't care about the Upstream lead time; you don't care about time spent in the *Ready* buffer, either. The problem here is a selection problem: Should you do it or not? It's not a scheduling problem anymore, because you can't calculate the cost of delay.

Takeaways

- Failure demand represents demand generated as a consequence of previous poor-quality deliverables.
- Failure demand in your process indicates low-quality work and prevents you from scaling.
- We measure abandoned work at Maturity Level 2, where the commitment point is still ambiguous, and the process is not clearly divided into Upstream and Downstream.
- In more mature Upstream systems, where demand is refutable, the items can be discarded in the decision-making process.
- Whenever we can determine the cost of delay of an item, we can measure the lead time in Upstream.
- When the cost of delay is unknown, we do not measure lead time but rather monitor the option expiry date.

6

Managing Invest and Shelve Options

In this chapter, we explore how to manage the options in Upstream Kanban:

- The value of information and how it affects decision making
- Managing Upstream work in progress (identifying the point of diminishing returns and applying the time guillotine)
- Managing shelved items and the *Pool of Ideas* (expiry date policy)

Figure 30 Balancing costs and benefits

Cost of Collecting Information

Collecting information doesn't come for free. As Donald G. Reinertsen says, "Product development inventory is physically and financially invisible. . . . But just because this inventory is invisible, doesn't mean it doesn't exist."[42] That inventory requires management, and the management of the Upstream inventory is mainly related to collecting information. There is a cost involved in gathering information to decrease uncertainty and decide whether to invest in an option or discard it, and it doesn't come for free. The cost may include:

- Capturing and processing information (cost of primary information):
  - direct search cost
  - indirect access cost
  - management cost
  - infrastructure cost
  - sunk cost
- Cost of secondary information (if we decide to buy information rather than collect it using our own resources)
- Potential cost of using information inefficiently when gathered:
  - information collected but not needed
  - information stored long after it's needed
  - information disseminated more widely than necessary
  - collecting the same information by more than one method
  - duplicating information

In this chapter, we explore techniques that support active management of the options to maximize decreasing uncertainty during the time they spend Upstream.

The Value of Information and the Value of Perfect Information

Decision makers may be offered a forecast of a future outcome (for example, a market research group may predict the forthcoming demand for a product). This forecast may turn out to be correct or incorrect.

> **Note:** In the Pareto-distributed domain, such as pharmaceutical research or movie production (script development), the information is impossible to value. In the Gaussian-distributed domain, you can always use the average value of information based on the known range of possible outcomes.

- With perfect information: The forecast of the future outcome is always a correct, 100 percent accurate prediction.
- With imperfect information: The forecast is usually correct but can be incorrect. Imperfect information is not as valuable as perfect information.[43]

---

[42] Reinertsen, 2009, page 55.
[43] https://kfknowledgebank.kaplan.co.uk/the-value-of-perfect-information-

the link in note 43 does not work

How much would you pay for (perfect) information?

The standard financial approach to this problem requires a probability and payoff analysis of expected outcomes. The formula for the Value of Perfect Information (VoPI) is:

> **Note**: Buying information helps to reduce uncertainty. The greater the uncertainty, the more we should be willing to pay to reduce it.

Expected Profit (Outcome) WITH the information LESS Expected Profit (Outcome) WITHOUT the information

Even though it is not common to perform this kind of detailed analysis when it comes to managing the Upstream Kanban, we still can assess the value of options and the value of the information those options carry.

Let's examine the following scenario:

We are undertaking an investment, and the forecast says that the revenue can be anywhere in the range of ten thousand to a million US dollars. Clearly, the uncertainty is very high.

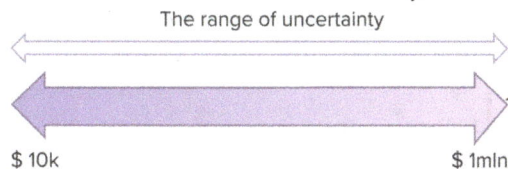

The range of uncertainty

$ 10k $ 1mln

Question: How much would we pay to narrow down this range ($100k–$1mln) (or 10x, one order of magnitude, effectively halving the uncertainty from 100x, or two orders of magnitude)?

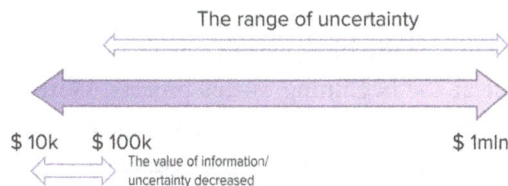

The range of uncertainty

$ 10k $ 100k $ 1mln
The value of information/
uncertainty decreased

The maximum amount we should pay for narrowing the range is ninety thousand dollars. That's the investment in collecting information. What does it mean from the Upstream perspective?

As we've already concluded, collecting information to decrease the uncertainty in the Upstream doesn't come for free. Hence, before we start applying techniques to manage Upstream options (time guillotine, the point of diminishing returns), we need to find a moment when continuing information collection becomes overinvestment. We shouldn't pay more than ninety thousand dollars for narrowing down the range to one hundred thousand to a million dollars. So, if we've already:

- paid a Business Analyst $90,000 for their work
- invested $90,000 in external market research

  or

- spent $90,000 worth of time on our internal research team

and we still don't know whether the minimal revenue will be $100,000, we have reached the point when we need to make a decision.

The Point of Diminishing Returns and Time Guillotine—Managing Upstream Work in Progress

Recognize that you would need to invest more (time, money, or resources) when the uncertainty is high, and much less when the uncertainty is low, as illustrated by the graph in Figure 31.[44]

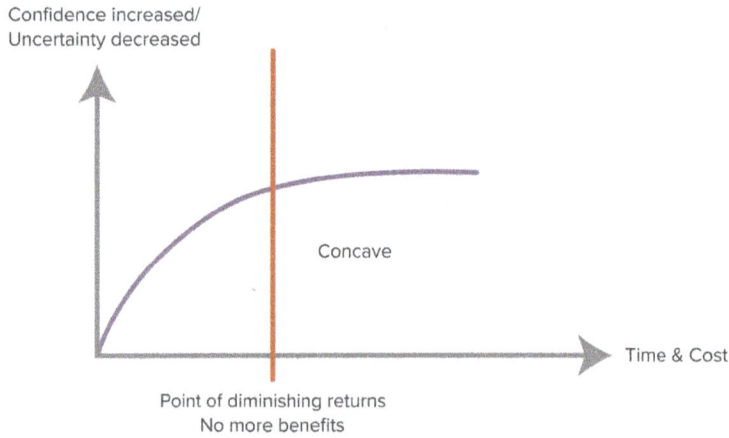

Figure 31 Identifying the point of diminishing returns

When we start collecting information, we can usually observe a fast confidence increase and uncertainty decrease at the beginning of this process. With time, it may slow down until the point where as time passes, we spend time and energy trying to collect even more information, but it doesn't happen, and our learning curve flattens. This moment, when we observe that there is no more information coming that would affect our decision-making process, is a signal that we have reached *the point of diminishing returns*. This pattern is known as *The Law of Diminishing Returns*, which is the "economic principle stating that as investment in a particular area increases, the rate of profit from that investment, after a certain point, cannot continue to increase if other variables remain at a constant. As investment continues past that point, the rate of return begins to decrease."[45] (See Figure 32.)

It is also true for the information collection in your Upstream process. At each step of this process, you will reach the point of diminishing returns. It is important to recognize it and make an appropriate decision to avoid stalled information collection and analysis.

A useful technique to manage Upstream work in progress is introducing the time guillotine. The main idea behind a time guillotine is to limit investment (the purchase price

---

[44] D. J Anderson and T. Bozheva (2021), p. 201.

[45] https://www.techtarget.com/searchcustomerexperience/definition/law-of-diminishing-returns

of the option) by limiting the time spent on it (i.e., the amount of work). So, the guillotine applies only when the item is in a work/information-discovery state.

Implementation Guidance

- Start with establishing the *Service Request Review* and the *Replenishment Meeting*.
- When reviewing the items in your Upstream system, ask the following questions:
  - What have we learned about this work item that is new since the last meeting?
  - Is this information significant enough to allow for deciding whether to promote or discard the work item?
- Whenever you observe that there is no new information arriving and that you have learned nothing (or very little) new since the last meeting or review, you need to decide what to do with this item:
  - You invest in it, as the risk is low, and you want to learn more about this option in the next phase or, if you have reached the *Ready to Commit* point, you want to commit to the work.
  - You discard it, as:
    ◦ the risk is high
    ◦ (and/or) the NPV is not favorable
    ◦ (and/or) you don't know enough, and you cannot invest more to increase the confidence.
  - You shelve it, as you don't know enough and you want to wait for changing business circumstances to increase your confidence later.

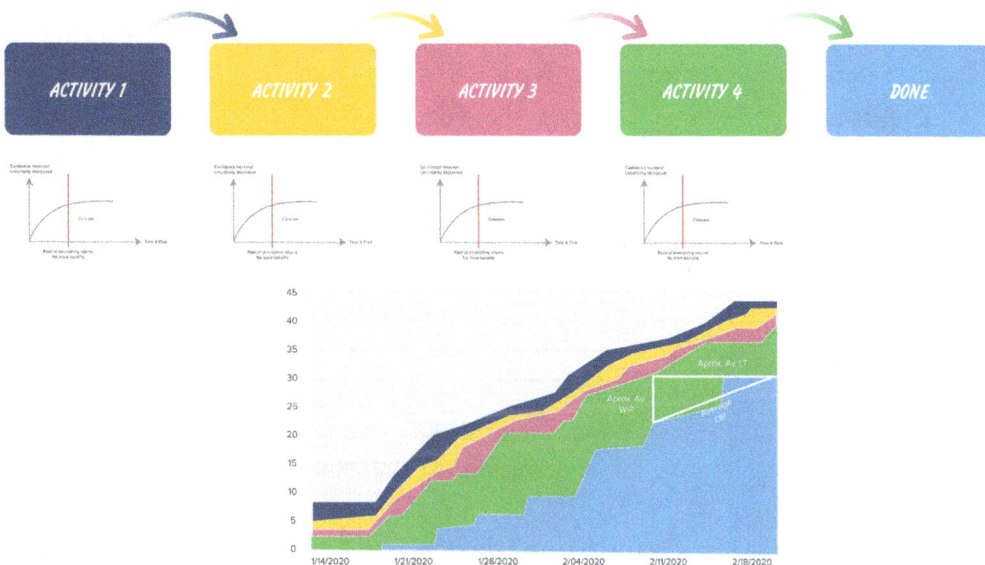

Figure 32 The point of diminishing returns indicates the moment for decision making.

Read more about Upstream cadences in Chapter 4, Upstream Feedback Mechanisms: Meetings and Reviews.

Expiry Date Policy—Managing Shelved Items

Ideas from the *Shelved Items* parking lot require special attention. Shelving options doesn't mean doing nothing with them.

The higher the uncertainty, the greater the option's value, and the more value in deferring commitment (read more about options valuation in Chapter 7, The Value of Deferred Commitment). However, the item's value must be at a significant premium above the re-

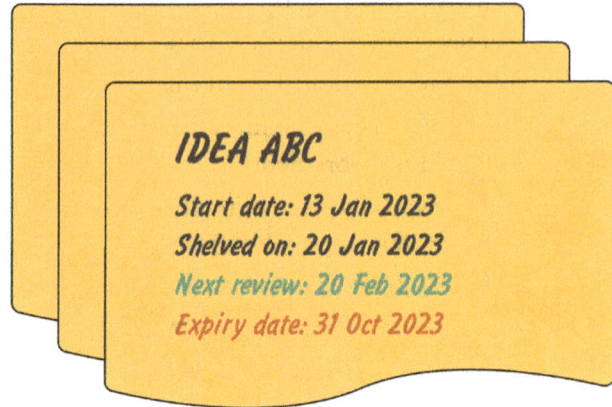

IDEA ABC

Start date: 13 Jan 2023
Shelved on: 20 Jan 2023
Next review: 20 Feb 2023
Expiry date: 31 Oct 2023

Figure 33   An example of a shelved item's ticket

quired investment cost. The option, which was shelved to wait, already had an initial investment embedded (remember value discounting over time). That's why we need to develop discipline and create a policy to ensure that our shelved options are not aging wastefully.

Implementation Guidance

- Even though knowledge work is intangible and invisible, think about it as if it occupies real space. Treat your *Shelved Items* parking lot as a fridge for your ideas. Apply a work-in-progress limit on it. When your "fridge" is full, you need to take something out before you put a new thing in its place. See Figure 33 for an example ticket.
- The same as food in your fridge, the shelved ideas have an expiry date, and you should keep an eye on it. Visualize it as an "Expiry date" time stamp on your ticket.
- Create an expiration policy: The ticket can stay in a particular column for a given period of time (x days/weeks/months). Add a "Next review" time stamp on your ticket to indicate when you need to make a follow-up decision about the option.
- At Service Request Review go through all shelved items. If an option reaches the decision point:
  - You move the idea back onto the board when a change in business circumstances indicates favorable outcomes and your shelved option was an option to delay. You start collecting information to decrease the risk further and manage it as a regular option from now on.
  - You discard the idea when a change in business circumstances indicates unfavorable outcomes and your shelved option was an option to delay.

- You are not able to make a decision yet—you shelve it further. Update the "Next review" time stamp and review the item in the next cycle.
- If you run out of time (you reached an expiry date)—the ticket is not an option anymore; you need to discard it.
- Create an explicit policy around decision points and expiration dates.

Expiry Date Policy—Managing the *Pool of Options*

A specific variant of an expiry date policy can be applied to the *Pool of Options* (or *Backlogs* at lower maturity levels). It is a commonly recurring pattern that the *Pool of Options* becomes an endless repository of all requests ever created in the company, and putting them all together reveals thousands of work items. To keep the process healthy, it is worth introducing a policy of a regular clean-up of the *Pool of Options* in your organization.

Supporting Cases

When introducing an expiry date policy, always consider the organizational context and the resistance the change may evoke. Whether you do it for the first or the second time, there is no definitive answer on how much time the policy should cut out. To illustrate, here are two different scenarios from organizations I had a chance to work with.

Scenario 1: Investment Bank

Starting a new role, I inherited a *Backlog* of more than 3,000 items. Some of them were a few years old and the employees who created the tickets were no longer working in that organization. Being asked to manage that backlog, I decided to limit the number of items to review. This was the knowledge inventory carrying too much cost, and it required reduction (read more about the negative aspects of that situation in Chapter 8, Upstream Inventory).

I filtered out all tickets that had not been updated in the previous three months and informed all ticket creators that their items would be closed if no action was taken. After that, I sent three more reminders at weekly intervals (which totaled a month of information shared in advance). Once we reached the deadline, I closed all tickets meeting the criteria; I also created a policy that covered a regular items' clean-up. In this way, I was able to trim down the number of *Backlog* items from more than 3,000 to a little more than 200.

Each idea has a "parent." And the parents love their "kids" unconditionally. If you close a ticket that is important to someone, this person will find you to tell you about it. Over the next few weeks, some of the ticket owners came to my desk asking, "How could you close my ticket?" That was a clear indication that the request probably should have remained open, so I simply reopened it.

> **Note:** Do not delete or remove tickets completely! Close them or change their status to indicate the nature of closure. This way you make them easy to find and reopen if necessary.

Scenario 2: FARA

When I worked as a consultant with FARA, one of the process improvements that we introduced was an expiry date policy for their *Backlogs*. At that time, all their *Backlogs* consisted of more than 30,000 open items, and they were impossible to manage. Some of the items, labeled as "critical defects," were twelve years old! Some other items covered upgrades to the systems and applications no longer in use.

However, due to the nature of their work, we couldn't introduce the aggressive expiry date, as I had done in other organizations. I realized that the whole idea of closing *Backlog* items made the Product team feel very uncomfortable, but I couldn't understand why. We discovered clearly that we could not start with an overly aggressive expiry date policy, as it created an impression of "losing information" among Product team members. Hence, we started with the safest possible policy, which said, "If a work item has not been updated in the previous four years, it becomes *Obsolete*." (*Obsolete* was a new label created for those items and indicated that the request was automatically closed when it met the policy criteria.)

Even though this solution left many items open, it still helped us to trim the number of open requests by half (which resulted in more than 15,000 closed items). What is more, we created a policy and respective automation in the system that closed all tickets meeting the established criteria at the end of each day. We also agreed to review the policy in three months to see if the expiry date period could be further shortened.

Takeaways

- Collecting information doesn't come for free. There is a cost involved in gathering information to decrease uncertainty and decide whether to invest in an option or discard it.

- Buying information helps to reduce uncertainty. The greater the uncertainty, the more we should be willing to pay to reduce it.

- The signal that we have reached the *point of diminishing returns* is the moment when we observe that there is no more incoming information that would affect our decision-making process.

- For shelved items, we need to develop discipline and create a policy to ensure that shelved options are not aging wastefully.

- If your ticket reaches an expiry date, it is not an option anymore, and you need to discard it.

7

The Value of Deferred Commitment

This chapter explores the topic of options valuation, in particular:

- The characteristics of options
- The value in deferring commitment
- Why deferring commitment is even more important when we observe high amounts of dark matter
- What types of commitment we have
- The types of uncertainty and how they affect the options valuation

Heuristic Reasoning

When talking about options valuation we don't mean "estimation"! We are not even aiming to achieve a high level of accuracy (not to mention precision). Real options valuation (ROV) may be difficult even for financial management analysts. Hence, it is often subject to heuristic analysis—a rule of thumb—that employs a pragmatic approach that is "good enough," allowing for flexibility and quick decision making in a complex, ever-changing environment, but it is based on sound financial criteria. The real options heuristic is simply recognizing the value embodied in the flexibility of choosing among alternatives even though their objective values cannot be mathematically determined with any degree of certainty.

There are certain characteristics of the real option that may provide us with guidance on what we should focus on when trying to assess the option's value:[46]

- ▶ Strike price—long options contracts are derivatives that give the holders the right but not the obligation to buy or sell an underlying security at some point in the future at a pre-specified price. This price is known as the option's strike price or exercise price.[47]
- ▶ Intrinsic value—the value of the option if we realize it today (the difference between market value and the strike price).
- ▶ Expiration date/time to expiry—the timeframe within which the business decision should be made, otherwise the value will be lost.
- ▶ Volatility—measures the level of risk in an investment; the higher the risk, the more expensive the option. The greater the volatility (range or swing in market price), the more valuable the option, and hence, to hold the option is worth paying more.

In their book *Commitment*, Chris Matts and Olav Maassen came up with a famous quote:

Options have value.
Options expire.
Never decide early unless you know why.[48]

Options Have Value

An option's value consists of an initial, nonrecoverable value and an intrinsic value (how much the option is worth today minus the strike price). The additional value of an option lies in its nature: It is your risk hedge against uncertainty. The higher the uncertainty, the greater the value of having the option.

If you decide to move from one country to another, one decision you need to make is where you will live. To do that, you start searching housing websites, asking friends and locals, and researching expat forums. Out of hundreds of options, you do not choose the first apartment that you click on. Rather, you narrow down your search using some criteria, and those are the *options you keep open* until you find the perfect apartment to live in. The value of those options lies not only in the time and effort you spend on managing them actively but most of all in the insurance it gives you so that choosing the wrong option too early becomes a binding constraint.

Options Expire

The expiration date or time to expiry is one of the most important characteristics of the options, as we don't keep them open in perpetuity. The option may either expire and no longer be available to exercise or expire and turn into an investment decision.

---

[46] https://www.investopedia.com/terms/r/realoption.asp
[47] https://www.investopedia.com/terms/s/strikeprice.asp
[48] Olav Maassen, Chris Matts, and Chris Geary, *Commitment* (Netherlands: Hathaway the Brake Publications, 2013).

Let's use the moving home and country example again. We checked a couple of different apartments that met the criteria, and now we need to decide on which one we would like to move into. What we found out is that:

- One apartment has already been rented by someone else—this option expired and is no longer available.
- We are moving next week, and we need to send our furniture across a few countries, so the only apartment we're left with becomes our place—this option expired and turned into an investment decision.

It's important to keep track of expiry criteria and conditions to avoid being left choiceless.

Never Decide Early Unless You Know Why

"Never decide early unless you know why" means deferring commitment. However, the apartment building is waterfront, brand new, located in a good neighborhood, close to transport, and an easy commute to work. Commit early, because you know why! It is very hard to imagine a better option becoming available at an affordable price.

Deferred Commitment

Deferring commitment[49] means not committing early, and instead waiting to decide on an item until more information is available. In a workflow, deferring commitment should reduce the amount of both aborted work and rework.

Deferring commitment implies that we should not decide until 1) we are sure we want to do something, and 2) we have as much information as possible that may affect the decision to do it or not.

However, every option has an expiry date, the point at which it is no longer viable to make a positive choice. The point just before that is referred to as *the last responsible moment* (LRM). While an option has an expiry date—a point at which it no longer makes sense to do it—we generally don't control the delivery date; instead, we control the decision to start an item. For the purposes of triage, selection, and kanban system replenishment, we have chosen to define the last responsible moment as the 50th percentile of the lead time before the desired delivery date (DDD). Beyond that point, we call it "irresponsibly late" (read more in Chapter 9, Plan Your Work! Cost of Delay, Triage, and Two-Phase Commitment).

Deferring commitment until the LRM comes with a trade-off: We probably need to treat that item with a higher class of service, probably a Fixed Date class of service; whereas if we start it earlier, we can trade some certainty and quality of information, and possibly rework or risk of aborting, for a lower class of service, such as Standard or Intangible. Read more in Chapter 9, Plan Your Work! Cost of Delay, Triage, and Two-Phase Commitment.

---

[49] D. J Anderson and T. Bozheva (2021), p. 185.

In the *Kanban for Design and Innovation* class, we use the following example to show the idea behind "Never commit early unless you know why."

In eighteen months, you must deliver a regulatory report that will take you four months to deploy; you know requirements will not change, and lead time is four months (no uncertainty about lead time).

Should you commit now? Or would you rather wait until the LRM?

Although this is an artificial case and the situation described above is very sterile (in reality, we don't have that high certainty about requirements or lead time), it illustrates what it means to *know the why*.

The usual answer from students is that they would wait until the LRM and commit to this work only then—due to the low/no uncertainty.

However, in this case, the low uncertainty is one of the *why*s. What is the full list, then?

- ▸ **The requirements uncertainty is low/none.** It means that the quality of information is high, the requirements will not change (even if we finish the work earlier), and the risk of rework is close to zero.
- ▸ **The lead time uncertainty is low/none.** We know exactly what time is required to deliver the work from the moment we start working on it.
- ▸ **This is a regulatory report.** It is required to run the business. The demand is irrefutable.
- ▸ **Treating the item with a lower class of service (Standard or Intangible) ensures a smooth flow.** In case any more urgent demand comes, it can be taken up and delivered immediately, and the regulatory report will not suffer. Committing right before the LRM, we either risk that we will not deliver the report on time, or we may miss another opportunity for potential revenue.
- ▸ **The option value is low due to low uncertainty and absolute certainty.** Hence, with very low option value and the knowledge that it must be done, we can and should commit early. The class of service it is given can then be flexible. This is, in and of itself, optionality, and it has value. If we defer and then commit at the LRM with Fixed Date class of service, then we've lost optionality. We've added unnecessary risk.

In general, we want to avoid committing early and encourage acquiring more knowledge to ensure that we commit to items that will not be aborted later. However, we do not want every item deferred until the LRM. Doing so would load too much risk onto our delivery capability.

Deferring commitment brings several benefits:

- It buys more time to refine understanding and improve certainty on requirements, design, and implementation.
- It reduces the chances for changes to the requested work or the promised deadline.
- It reduces the probability of aborting requested work items.
- It reduces the overhead for managing modifications, reprioritization, and replanning.
- It reduces the coordination cost for holding meetings.

Dark Matter

According to CERN, an international organization that studies the universe, dark matter is an invisible mass that might account for as much as 95 percent of the total mass of the universe.[50] Hence, whatever we see accounts for only 5 percent.

---

[50] https://home.cern/science/physics/dark-matter

The dark matter of requirements can be the case for most of the processes facing the Upstream challenge, as dark matter lies in the gap between perfect information and the imperfect information we currently hold.

Mature teams working in well-understood domains produce less dark matter. Immature teams may find that the lack of information about what the work requires may lead to 100 to 200 percent more work than they had planned.

When we make progress with imperfect information, we should anticipate greater amounts of dark matter. That's one more reason why deferring commitment is beneficial—it reduces the mass of the dark matter of information.

Types of Commitment

In this section, we examine the different types of commitment:

- Synchronous, asynchronous
- Reversible, non-reversible
- Symmetrical, asymmetrical

Synchronous and Asynchronous Commitment

We talk about **synchronous commitment** when:

- Both parties (Upstream and Downstream) are available.
- It happens at one point in time (e.g., Replenishment Meeting).
- There is common agreement and a real or virtual handshake.

Committed work is pulled to the *Delivery in Progress* column directly from *Discovery Done* during the Replenishment Meeting.

This is a mutual commitment to "what" and "when." The commitment is irreversible (Figure 34).

Asynchronous commitment is when we have a separation in the decision points (in time and, possibly, space). Upstream commits and creates a "committed buffer" from which Downstream has to commit. Typically, the Downstream commitment is deferred until "now," so "when" is the point in time of the Downstream commitment. There isn't much concept of scheduling; instead, an item is pulled when Downstream is ready to commit.

This is comparable to a short-order restaurant system. The server takes the customer's commitment, pins the request on the kitchen's rotating wheel, and the kitchen pulls the request when they have the capacity to cook. There is a buffer between the order-taking process and the order-fulfillment process.

- Both parties (Upstream and Downstream) are available.
- It happens at different points in time.
- There is common agreement but no handshake (agreement is based on policies).

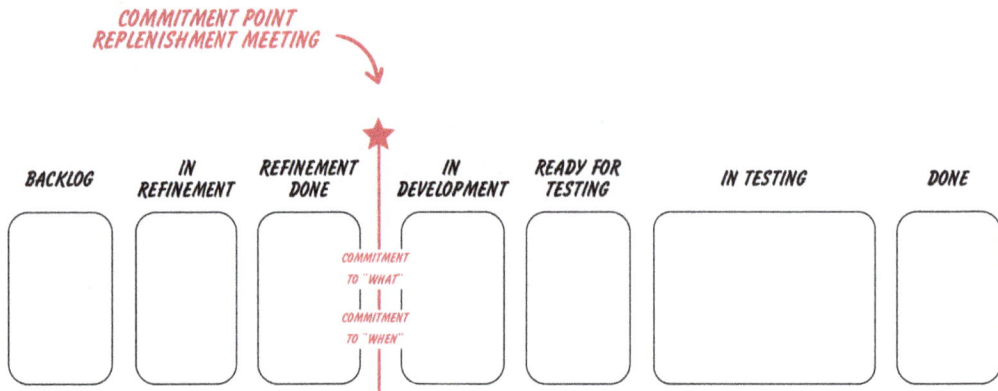

Figure 34 Synchronous commitment

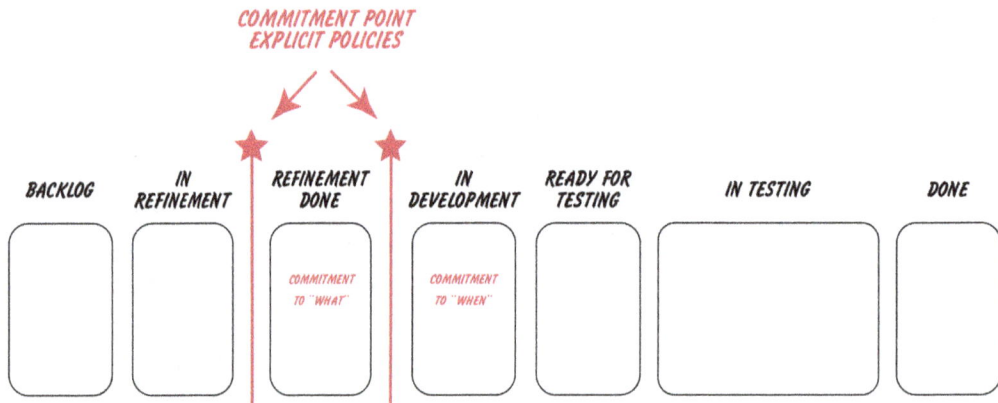

Figure 35 Asynchronous commitment

The Upstream part of the process commits to what needs to be delivered, but the work item stays in the Upstream (*Refinement Done* in Figure 35). The commitment is still reversible.

The Downstream part pulls the work item whenever there is available capacity. When the work is pulled from the Upstream system (*Refinement Done*) to the Downstream system (*In Development*), it is considered as committed to start (commitment to "when"). The commitment becomes irreversible.

Reversible and Irreversible Commitment

The idea behind deferred commitment is that we postpone deciding to minimize the risk of the unknown. We should keep our decisions reversible and flexible and avoid locking them down too early when the uncertainty is still high.

Formulate an explicit policy that determines when the commitment becomes irreversible. Answer the following questions:

- Is there an option to back out of the commitment?
- What are the consequences of reversing a commitment?

This decision is similar in nature to pulling a product from the market (read more in the section Cost of Quality later in this chapter), although the consequences are not as heavy. It is still, however, worth understanding when our commitment becomes irreversible to avoid aborting the work after it has been committed.

Symmetrical and Asymmetrical Commitment

The distinction between symmetrical and asymmetrical commitment is all about risk placement.

On the hugely popular website booking.com, looking at hotel prices, for example, we can observe how a hotel balances risk. Let's take the example in Figure 36: four nights for a hotel in Bilbao between February 25th and 29th.

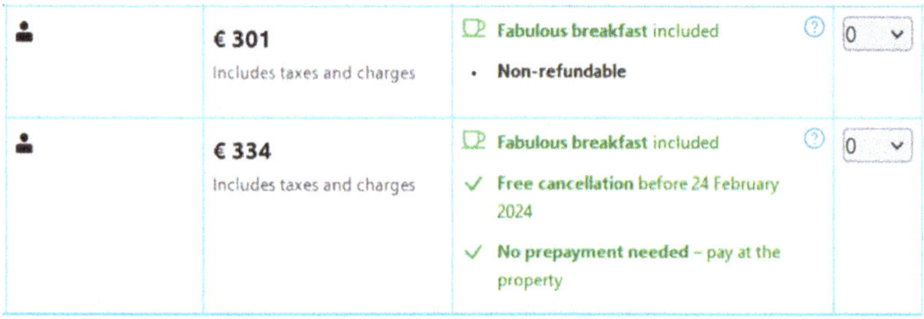

	€ 301 Includes taxes and charges	⏏ Fabulous breakfast included • Non-refundable	⑦ 0 ⌄
	€ 334 Includes taxes and charges	⏏ Fabulous breakfast included ✓ Free cancellation before 24 February 2024 ✓ No prepayment needed – pay at the property	⑦ 0 ⌄

Figure 36   Options for a room at booking.com

- The first option is cheaper—but non-refundable. It means that while booking your stay, you risk being a no-show, as you have to pay immediately to complete the reservation. The hotel offers you a lower price in exchange for lowering their risk. You choose this option when your uncertainty is low—you know that you can invest and commit early.
- You pay more for the other option, but you can defer commitment until later. First, you can cancel for free until almost the last responsible moment (one day before arrival), and second, you pay at the property. The hotel takes the risk of your not showing up, but as a recompense, you are charged more.

This example represents a mature approach to commitment, where the risk is distributed among both parties of the transaction.

At the workplace, we observe a shift to symmetrical commitment at Maturity Level 3 through fully functioning Replenishment Meetings, when there is a clearly defined commitment point, and risk is now better shared by the Upstream business organization and the Downstream delivery organization.[51]

At lower maturity levels, the members of the delivery organization are not partners in delivering the best outcome. Rather, they are treated like vendors, in a low-trust, low-maturity fashion, with the implication that their services can be easily replaced with an external vendor. Risk management is asymmetrical, and the organization is vulnerable to abusive and bullying behavior—this leads to very poor decision making and a delivery organization that is set up to fail and take the blame for failing.[52]

When a delivery organization is made to carry a disproportionate amount of risk, they tend to hedge that risk by building contingency into their time and budget estimates. This leads to suboptimal decision making—starting too soon (due to schedule padding) or spending too much money (due to padding cost estimates) or cutting scope or abandoning a project altogether (due to excessive cost estimates).

To verify whether we observe symmetrical or asymmetrical commitment, ask the following questions:

- Do both parties commit in the same way or not?
- Is there a balance of risk and responsibility for the decisions made?
- Are there policies in place establishing risk sharing between the Upstream and the Downstream?

Types of Uncertainty and How They Affect Options Valuation[53]

"Managerial flexibility has value in the context of uncertain projects, as management can repeatedly gather information about uncertain projects and market characteristics and, based on this information, change their course of action. This value is now well accepted and referred to as 'real option value.'"[54]

To understand how the value of an option is affected by uncertainty, Arnd Huchzermeier and Christoph Loch identify five types of uncertainty: market payoff variability, schedule variability, budget variability, performance variability, and market requirement variability.

Market Payoff Variability

The market payoff (e.g., price and sales forecast) depends on uncontrollable factors such as competitor moves, demographic changes, substitute products, and so on. A wider spread of market payoff uncertainty increases options valuations, as illustrated in Figure 37. You should spend more time and/or money to acquire options.

---

[51] D. J Anderson and T. Bozheva (2021), p. 72.
[52] D. J Anderson and T. Bozheva (2021), p. 349.
[53] David J Anderson and Alexei Zheglov, "Enterprise Services Planning," presented at Lean Kanban Russia, October 2016.
[54] Arnd Huchzermeier and Christoph H. Loch, "Project Management under Risk" (INSEAD, 1998).

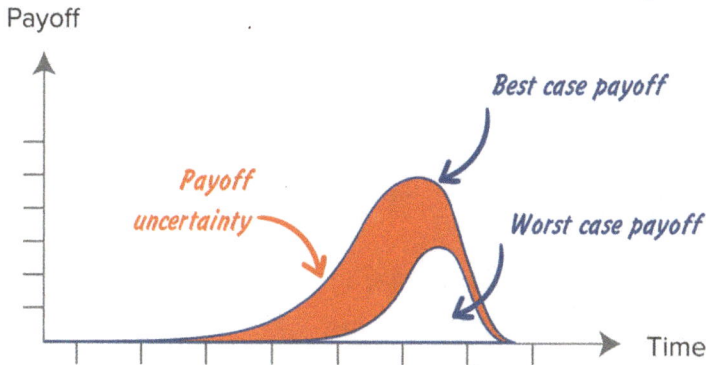

Figure 37 Diagram illustrating the market payoff variability

Schedule Variability[55]

The project may finish unpredictably ahead of or behind schedule. In the latter case, reduced market payoffs (in terms of market share or prices) may result, as empirical work shows. The likely payoff functions can be graphed.

Payoff Function Shape versus Schedule

A **front-loaded** payoff function, as depicted in Figure 38, is highly sensitive to schedule uncertainty; options are worth more. Spending more money and time to develop options is appropriate.

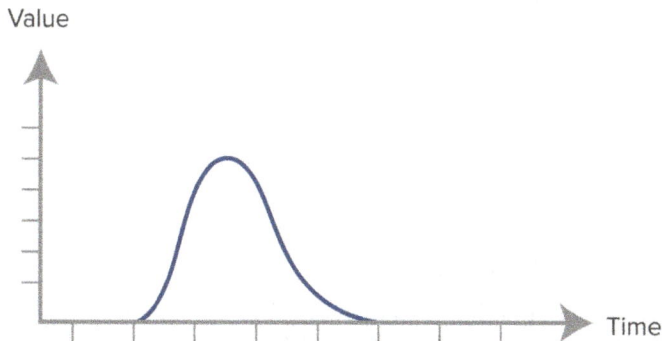

Figure 38 Front-loaded payoff function

A **bell curve** payoff function, as shown in Figure 39, is less sensitive to schedule uncertainty; options have lower value, but investing some time and money is still appropriate.

---

[55] See also Triage Tables in D. J Anderson and T. Bozheva (2021), p. 403.

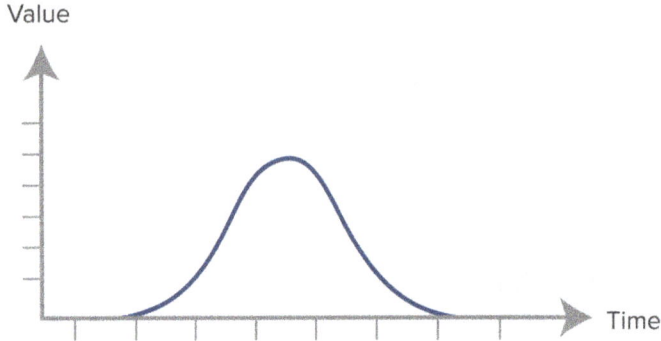

Figure 39 Bell curve payoff function

A **back-loaded** payoff function (Figure 40) has the lowest sensitivity to schedule uncertainty; try to avoid spending much time or money developing the option before commitment.

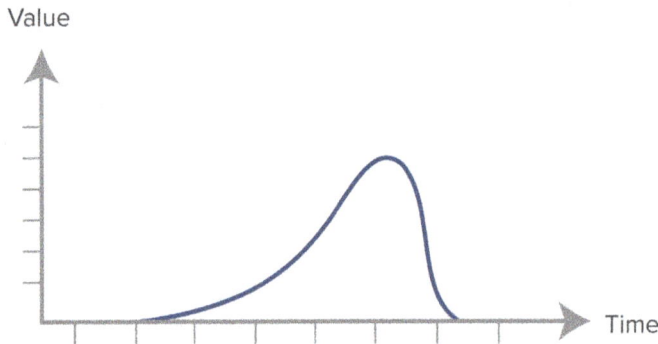

Figure 40 Back-loaded payoff function

Shelf Life

A product's shelf life also affects an option's value.

- **Ultra-short** (less than one product cycle): extremely sensitive to schedule uncertainty; options are tremendously valuable. Spending a lot of money and time to develop options is necessary, and dedicating people and resources may be essential.
- **Short** (one product cycle): highly sensitive to schedule uncertainty; options are worth more. Spending more money and time to develop options is appropriate.
- **Medium** (two to four product cycles): less sensitive to schedule uncertainty; options have lower value, but some time and money invested is still appropriate.

- **Long** (five to nine product cycles): very low sensitivity to schedule uncertainty, so try to avoid spending much time or money developing the option before commitment.
- **Very long** (ten or more product cycles): extremely low sensitivity to schedule uncertainty, so try to avoid spending much time or money developing the option before commitment.

Budget Variability

Budget variability acknowledges that the running development costs of a project are not entirely foreseeable. Budget overruns are common and, less frequently, underbudget completion also occurs.

Being over budget tends to happen because of dark matter. Coming in under budget can happen when a cheaper, better solution is found for the delivery or implementation of specific features of the project or product. The arrival of hidden requirement information can cause budget overrun, while the arrival of hidden implementation or delivery solution information can result in budget savings. For example, a new tool that automates a previously manual process means that budgets based on historical costs are no longer valid—essentially, a small punctuation point happened, and historical data is invalid.

- Work items that have fixed costs, fixed burn rates, and fixed delivery dates have low budget uncertainty. You can commit early and spend less on options.
- Spending more to learn more about an option may reduce budget uncertainty, but it is subject to a law of diminishing returns. It is good to do initially, but each improvement in budget certainty decreases the option valuation; hence, further investment in reducing budget uncertainty may be wasteful.

Performance Variability

Performance variability refers to uncertainty in the performance of the product being developed. Initially targeted performance often cannot be fully achieved, as trade-offs must be resolved among multiple technical criteria, which together determine performance in the customer's eye. The greater the technical novelty of a product, the higher this uncertainty.

- Quality uncertainty decreases option values.
- If our delivery quality is uncertain, we should not spend a lot developing ideas prior to commitment.

Read more about this in the section Cost of Quality, which follows.

Market Requirement Variability

Market requirement variability means that there is uncertainty about the performance level required by the market. Performance targets for a product are often only imperfectly known, especially for conceptually new products.

- Requirements uncertainty reduces an option's value.
- When we are not sure of the specification, we shouldn't spend a lot of time or money defining the specification.
- Instead, we should have lots of cheap or free options and explore them with iterative incremental commitment.
- It's better to run lots of cheap experiments and amplify the ones that show promise.
- With lower uncertainty, fewer higher-value options are optimal, and it is acceptable to spend more money and time to define precise requirements. Legacy system replacement would be a good example.

Cost of Quality

There is one more cost related to the way we manage options, namely, the cost of quality.

We all agree that quality is important, but quality has its cost, and it depends on your product or service, your market, and how much you can afford to pay for it.

Talking about the cost of quality, we distinguish two groups:

- Cost of operating a quality management system (QMS), also called the cost of conformance (to quality):
  - Inspection or appraisal cost (testing, peer reviews, customer reviews, and so on)
  - Prevention cost (quality training, supplier surveys, quality planning, improvements)
- Cost of failure (the cost of nonconformance to quality):
  - Internal failure cost: scrap, rework, reinspection, fixing defects (after the request was completed but before it was released to the customer)
  - External failure cost: complaint processing, warranty claims, product recalls; but also damage to your reputation, which can result in loss of value of your market shares

Both internal and external failure come back as failure demand in your Kanban system.

Cost of Delay due to Ensuring Quality

While it is not commonly discussed in the literature, the flip side of failure delay is the cost of delay due to ensuring quality. What is important—high quality—doesn't mean "the best." It means what is proper under the circumstances. A high-quality product produced to high standards will give good service.

You need to balance the cost of conformance to quality with the cost of delay. To do that, use options thinking.

Recognize that the cost of rework can be traded for option pricing. Ask yourself whether an option is reversible and what the consequences are of reversing it. For example, a marriage can be reversed with a divorce, but this comes with a cost.

Releasing a product to the market that must be recalled or reworked and replaced may have unacceptable consequences, in which case it is worth paying more for the option up front (Figure 41). On the other hand, if the consequences of problems in the market are perceived as minor, then the option price would be lower (Figure 42); it would be better to proceed and course-correct later.[56]

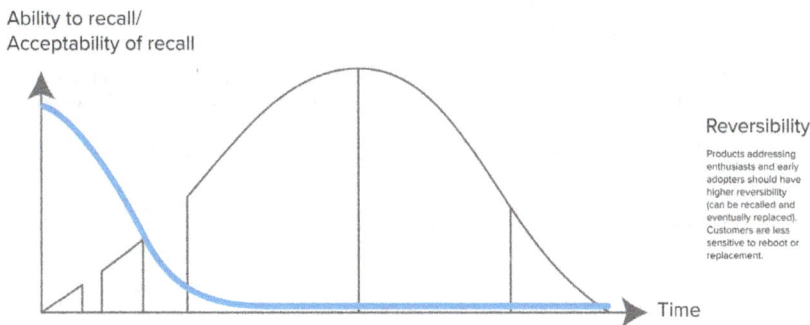

Figure 41 Product reversibility is higher in the early stages of the product life

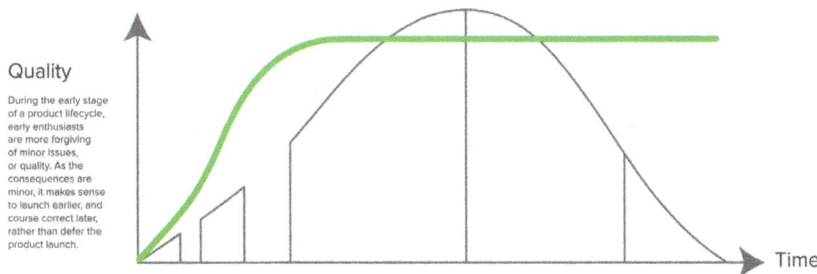

Figure 42 Quality level expectations differ among different markets

Before we decide to release our product, we need to ask the following questions:
- Is our option reversible?
- What are the consequences of reversing it?
- Are these consequences perceived as minor or unacceptable?
- What is our customers' tolerance of potential mistakes and problems?

[56] D. J Anderson and T. Bozheva (2021), p. 201.

- How much should we pay up front for our option to avoid unacceptable consequences?
- How much of the cost of conformance to quality should we bear to avoid unacceptable consequences?

Summary

Table 2 presents a summary of the types of uncertainty and their effect on an option's valuation and the possible Upstream decision.

Table 2 Summary of Types of Uncertainty and Their Effects

Category of Uncertainty	Scale of Uncertainty	Effect on Upstream Decision and Options' Valuations
Market payoff variability	A wide spread of market payoff	Increases options' valuations. You should spend more time/money to acquire options.
	Low spread of market payoff	Low options value. You should avoid spending time/money to acquire options.
Schedule variability	Front-loaded	Highly sensitive to schedule uncertainty; options are worth more. Spending more money and time to develop options is appropriate.
	Bell curve	Less sensitive to schedule uncertainty; options have lower value, but some time and money invested is still appropriate.
	Back-loaded	Lowest sensitivity to schedule uncertainty; try to avoid spending much time or money developing the option before commitment.
Shelf life	Ultra-short	Extremely sensitive to schedule uncertainty; options are tremendously valuable. Spending a lot of money and time to develop options is necessary, and dedicating people and resources may be essential.
	Short	Highly sensitive to schedule uncertainty; options are worth more. Spending more money and time to develop options is appropriate.
	Medium	Less sensitive to schedule uncertainty; options have lower value, but some time and money invested is still appropriate.

Table 2 Summary of Types of Uncertainty and Their Effects (*continued*)

Category of Uncertainty	Scale of Uncertainty	Effect on Upstream Decision and Options' Valuations
Shelf life	Long	Very low sensitivity to schedule uncertainty, so try to avoid spending much time or money developing the option before commitment.
	Very long	Extremely low sensitivity to schedule uncertainty, so try to avoid spending much time or money developing the option before commitment.
Budget variability	Low budget uncertainty (fixed cost, fixed burn rate, and fixed delivery date)	Commit early and spend less on options.
	High budget uncertainty (unknown requirements, dark matter)	Defer commitment and invest in collecting information before commitment.
Performance variability	High ability to recall/acceptability of recall (usually early markets)	Commit early, test options, and collect information directly from early or beta-testers.
	No/low ability to recall/acceptability of recall (usually mature markets)	Defer commitment and invest in collecting information before commitment.
Requirements variability	High requirements uncertainty	Avoid spending a lot of time or money defining the specification. Generate lots of cheap or free options and explore them with iterative incremental commitment.
	Low requirements uncertainty	Fewer higher-value options are optimal, and it is acceptable to spend more money and time to define precise requirements.

Takeaways

- When doing the options valuation, we are not looking for accuracy. Rather, we use the rule of thumb, allowing for flexibility and quick decision making in a complex, ever-changing environment.
- Options have value. Options expire. Never decide early unless you know why.
- We refer to "dark matter" as the gap between perfect information and the imperfect information we currently hold.
- Synchronous and asynchronous commitment refer mostly to the point in time when the commitment is being made.
- If our commitment is irreversible, we should delay decision making as long as possible and use the time to collect information that will amplify our choice.
- The distinction between symmetrical and asymmetrical commitment is all about risk placement.
- We identify the following types of uncertainty: market payoff variability, schedule variability, shelf life, budget variability, performance variability, and market requirements variability.

8

Upstream Inventory

Since the inventory in a product development process is not a physical object but information, it is virtually invisible.

—Donald R. Reinertsen

The Cost of Inventory

A key reason to hold inventory is to buffer delivery time and uncertainty in the deliveries' arrival. The concept of zero inventory in manufacturing requires near-continuous, uninterrupted, completely predictable, no-variability delivery/arrival stock when and where you need it. It requires a perfectly smooth flow. That is possible only when working with consistent components that require almost identical effort and processing time from one item to the next. That is impossible to achieve with knowledge work. For example, if you receive a new batch of information only once per month, you need to hold one month of inventory and buffer some uncertainty in demand. So, on average, you hold more than one month of information inventory. Even if you can pull information on demand, the amount of time and effort required to discover the information will vary dramatically from one request to another, even when the requests are of a similar type.

Unlike tangible goods, knowledge work inventory does not use physical space, but it does cost money, and hence, it has a carrying cost in addition to other costs identified so far for tangible inventory.

Let's look at them in turn.

Five Types of Inventory Costs[57]

1. Ordering Costs
   - *For tangible goods*: Ordering costs include payroll taxes, benefits and the wages of the procurement department, labor costs, and so on. These costs are typically included in an overhead cost pool and allocated to the number of units produced in each period.
   - *For knowledge work:*
     - Searching for information
     - Ordering (buying) information research (e.g., market research)
     - Transportation costs of information (passing information through a meeting or email)
     - Cost of electronic data interchange

2. Inventory Holding Costs
   - *For tangible goods*: This is simply the amount of rent a business pays for the storage area where they hold the inventory. This can be either the direct rent the company pays for all the warehouse space it uses or a percentage of the total rent of the office area utilized for storing inventory.
   - *For knowledge work*:
     - The cost of storing information (e.g., hard drives, servers)
     - Risk prevention costs (securing information from theft or altering)
     - Opportunity cost—money invested in information storage

3. Shortage Costs
   - *For tangible goods:* Shortage costs, also known as stock-out costs, occur when items become out of stock.
   - *For knowledge work*:
     - Idle resources
     - Disrupted process costs

4. Spoilage Costs
   - *For tangible goods:* Perishable inventory stock can rot or spoil if not sold in time (like bananas in the fruit market), so controlling inventory to prevent spoilage is essential. Products that expire are a concern for many industries. Industries such as food and beverage, pharmaceutical, healthcare, and cosmetics are affected by the expiration and use-by dates of their products.

---

[57] https://www.deskera.com/blog/inventory-cost/

- *For knowledge work:* Information is perishable—it degrades over time.[58] What you know to be true now may not be true in the future. What you know now, with clarity and precision, may fade to something vague and imprecise later. Some information is perishable more quickly than others—the weather forecast, for example. The spoilage cost may include the cost of rework or the need for collecting new information again.

  The faster we can turn a request or an idea into a finished product, the lower our carrying cost of inventory.

5. Inventory Carrying Costs
    - *For tangible goods*: This is the lesser-known aspect of inventory cost. Inventory carrying costs refer to the amount of interest a business does not earn on the value of unsold stock sitting in warehouses.
    - *For knowledge work*: Carrying the cost of inventory—the cost of keeping it "alive," continuing to think about it, talk about it, and discuss it in the meetings—all the coordination costs that are related to intangible inventory management. The bigger the inventory, the more carrying cost it has.

    Read more about techniques for inventory management in Chapter 6, Managing Invest and Shelve Options.

Optimal Batch Size

In classes I teach, one of the most popular questions I get (right after "What size should the ticket be?") is "What should the size of the batch be?"

Before answering this question, it's worth bringing up two factors that help us to decide the size of the batch: transaction cost and (inventory) holding cost. To illustrate this concept, let's look at the example used by Don Reinertsen.[59]

You like eating eggs, and you eat one egg a day. Before you can start preparing and eating the eggs, first you need to:

- Go to the store and buy eggs—these represent the transaction cost.
- Store the eggs in the fridge—this is holding cost.

There are two scenarios representing two extremes of the scale:

1. You buy an inventory of eggs for a year:
   - Your transaction cost is very low (you shop once per year).
   - Your holding cost is very high (you need a huge fridge, electricity to keep it going, and so on).
   - Additionally, part of your inventory will spoil.

---

58 D. Anderson, "20 Years of Flow," presented at Kanban Brazil, 2023.
59 https://www.youtube.com/watch?v=UOt0QW99-kE

2. You buy one egg every day, and you eat one egg every day (the idea of a one-piece flow associated with just-in-time production):
   - Your transaction cost is very high (you need to shop every day).
   - Your holding cost is very low, close to zero.

Neither of these scenarios is optimal; you need to find a proper batch size of eggs that keeps both transaction and holding costs at the lowest possible level. This is illustrated by Don Reinertsen's graph in Figure 43.

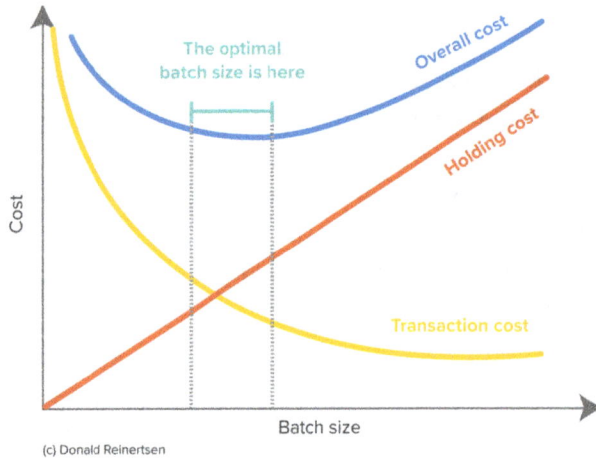

(c) Donald Reinertsen

Figure 43 Optimal batch size as depicted by Donald Reinertsen

How do you come up with the optimal batch size for your process? As Don Reinertsen says,[60] it's safe to assume that we won't be able to find an accurate optimal batch size. The only safe and useful heuristic is to assume that whatever we calculate will be lower than the actual batch size. As a minimum, you should aim for a batch size that is 30 percent smaller than whatever you think is optimal.

We must test the response of the process to the batch size changes that we are introducing by measuring the results (predictability of the process, lead times, efficiency, and so on).

> **Note:** Batch size reduction is reversible. Reducing the size by too much normally can be easily fixed by adding more scope. On the contrary, if we add too many work items to our batch, it can be difficult to reverse it. What is more, the reversibility element of smaller batches allows us to quickly modify the batch size in response to changing conditions.

How to Manage Large Backlogs

Our inventory is bits on a disk drive, and we have very big disk drives here at HP!

—Engineering manager at Hewlett-Packard[61]

---

[60] D. Reinertsen (2009), p. 122–125.
[61] D. Reinertsen (2009), p. 56.

The backlogs of work, especially virtual—when you are not limited by the size of your physical board—can be huge. In a rather small company of 300 developers, the backlog of work was more than 30,000 items and growing.

The challenges with a large set of choices are:
- What should we start next?
- In which order must we do things?
- How many things can we do in parallel?
- What might be the impact if an item takes longer than we expected?
- Do we need to repeatedly stack-rank them all?

Due to the perishable nature of information and intangibility of knowledge work, the priority of items keeps changing frequently, and the constant reprioritization becomes a huge overhead cost. On top of that, the definition of what priority really means often remains unclear.

Prioritization of a Kanban Backlog

"Priority" means "how the request is treated." And we can look at it from different angles:
- Sequencing—does this go before or after that?
- Scheduling—when should it start?
- Selection—which should we pick to start?
- Class of service—how do we treat this in relation to other requests, or the queuing discipline used to handle and advance the request.

What Is a Kanban Backlog?

Going back to the industrial tangible-goods era, a backlog is synonymous with an order book; that is, we have an order book of five naval vessels for the shipyard; it could equally be we have a backlog of five naval vessels for the shipyard. The orders are firm and are committed, perhaps even scheduled.

In that sense, backlog implies already committed. So, backlogs are appropriate for:
- an organization at Maturity Level 2 or below, with early commitment;
- project and portfolio management where the scope of the project is already committed because the project is committed;
- work that is spawned as the child of a parent item that is already committed (e.g., a backlog of photo shoot requests for the photographer because the ad agency has committed to a collection of campaigns for various clients);
  or
- a portfolio where projects are committed in an annual planning cycle.

A backlog is **not appropriate** where deferred commitment is an advantage, or when we want to actively manage options in an Upstream discovery process.

A backlog implies (big) batch transfers, whereas a pool of options implies individual commitment and single-piece flow. Hence, rather than using *Kanban Backlog*, we instead refer to the *Pool of Options*, not big-batch transfers.

Don't Prioritize!

Easier said than done! What does it mean not to prioritize? How can we know what we should do next if we don't prioritize?

Here is the expanded advice:

Don't prioritize! Filter instead!

How do we filter large backlogs? Let's use an example of an enormous pool of possibilities: flight tickets.

If you want to travel from one place to another, you don't simply put "flight" in your search engine. Without even noticing it, you start the filtering procedure.

Filter #1: *The starting point and destination*: Bilbao to Munich. Too many results (thousands of them!). We need to filter more.

Filter #2: *Dates of travel*: Using the filter shown in Figure 44, we pared it down to more than ninety results. Not bad compared to the initial set of results, but still too many.

Filter #3: *One stop or fewer* (Figure 45): fifty-three results; better still.

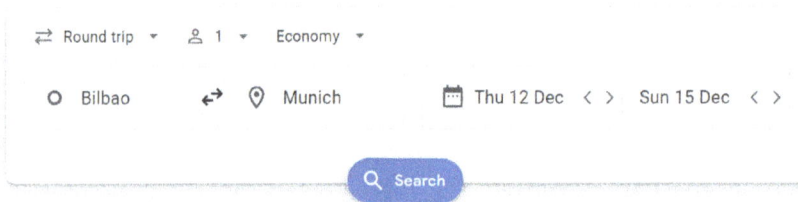

Figure 44 Filter by travel dates

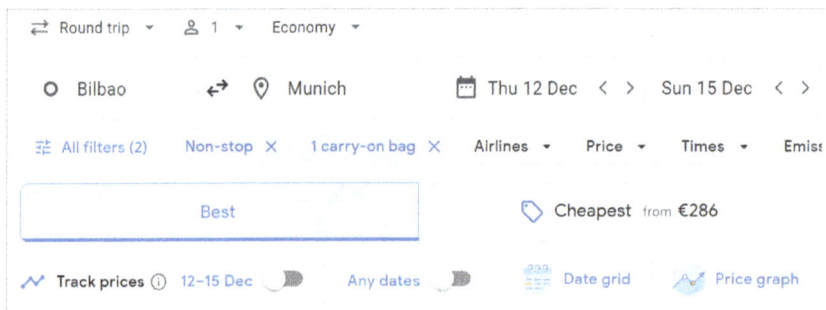

Figure 45 Filter by number of stops

We can keep doing this until we winnow the options to a manageable number that we are able to discuss and decide about. If your screen shows four, five, or seven options, you can start applying specific qualification criteria based on the risk assessment.[62]

Implementation Guidance

If you are dealing with large backlogs, it is crucial to filter before you start putting things in any order or scheduling them.

When searching for flights, we can select the feature "all (available) filters." You need to create an equivalent of this for your Kanban backlog: a set of criteria that will help you filter the number of options requiring an immediate discussion and decision on whether or not they should be pulled next.

Additionally, an explicit policy should be put in place to support the decision-making process.

Supporting Case: BestDay

At BestDay, the option filtering was done using three questions:

- Is this viable? That is, is it worth our time and effort?
- Is this something that our customers need?
- Can we do it?

The questions supported understanding the trade-off between the risk and the value of the items waiting to be pulled into the Upstream process. Their filtering board is shown in Figure 46.

Figure 46 Options filtering technique used by BestDay

---

[62] https://www.youtube.com/watch?v=PP0-oZBgE94&t=2005s

Takeaways

- ◆ A key reason to hold inventory is to buffer delivery time and uncertainty in the arrival of deliveries.
- ◆ We identify five types of inventory costs: ordering cost, inventory holding cost, shortage cost, spoilage cost, and inventory carrying cost.
- ◆ The factors that help us to decide on the batch size are transaction cost and (inventory) holding cost.
- ◆ Normally, reducing the batch size by too much can be easily fixed by adding more scope. On the contrary, if we add too many work items to our batch, it can be difficult to reverse it.
- ◆ A backlog implies (big) batch transfers, whereas a pool of options implies individual commitment and single-piece flow.
- ◆ Whenever we refer to a *Kanban Backlog*, we mean the *Pool of Options* rather than big batch transfers.
- ◆ Don't prioritize the backlog! Filter it!

9

Plan Your Work! Cost of Delay, Triage, and Two-Phase Commitment

There are a few evergreen planning-related questions:

- "What is the progress?"
- "When will it be done?"
- And finally, "When should this be delivered?"

We should stop using these questions, however, and replace them with: "When should we start to deliver on time?" Although the change may sound trivial or even insignificant, it helps you focus on what is important in planning and considers the non-homogeneous nature of our work.

When talking about proper, fact-based planning, we need to consider elements like:

- Classes of service based on the Cost of Delay
- Two commitment points
- Triage
- Start-date ranges
- Customer expectation

We look at them in turn.

Cost of Delay

Cost of Delay (CoD) is the relation between time and value expressed monetarily. It is the value we lose by delaying the introduction of our product or service. The numerator in the cost-of-delay equation is always a financial measure, and the denominator is the unit of time (e.g., $100/day, $1,000/month).[63]

The details of the Cost of Delay are described in the famous book by Donald G. Reinertsen, *The Principles of Product Development Flow*.[64]

The Cost of Delay has two aspects:

1. The actual cost:
   - The fine we need to pay for breaking the contract Service Level Agreements and
   - The penalty for missing a due date for in regulatory reporting or something similar
2. The opportunity cost—potential revenue we lost because we didn't deliver on time, or we delayed delivery.

Figure 47 shows an example of the impact the delay in a marketing campaign has on the room booking at a hotel awaiting their clients at Eastertime. We can see what happens if the desired delivery date is missed. The graph introduces a delay and shows the consequential actual sales of hotel rooms. The remaining red area on the chart illustrates the difference between the anticipated results and the actual results. This red area in the aggregate is known as the delay cost.

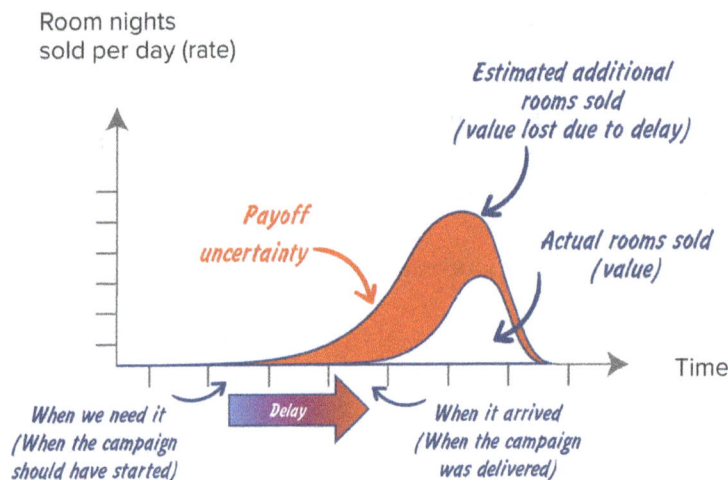

Figure 47 Value Acquisition Life Cycle function showing a delay in delivery

---

[63] Donald Reinertsen uses the term Delay Cost function to show the shape of the delay cost over time. Cost of Delay means the rate of delay cost over a specific, usually immediate, period of time. In other words, Cost of Delay is the derivative of Delay Cost. This is ambiguous in English and extremely problematic in translation. There are precedents, though. We use capacity to refer to the rate at which water flows through a pipe, but it can also mean the volume of a container of water, so a hot water tank can have a capacity of 250 liters, while a heat pump's capacity might be 100 liters per hour. If you use up all the hot water, it takes 2.5 hours to refill the tank.

[64] D. Reinertsen (2009).

The idea of the cost of delay in Upstream Kanban is widely used and can help establish the rules behind the lead time measurement (see more in Chapter 5, Upstream Metrics), as well as support making planning and triage decisions.

The Wedding Cake Problem

CoD is a concept that we face every day, and we can use it quickly to assess the risk of our plans and take appropriate risk-hedging actions. To illustrate that, we use the example of a wedding cake.

A wedding is an event that (in the majority of cases) is planned months (or even years) in advance and involves a lot of preparation, sub-projects, and dependencies. In short, it has tons of risks and uncertainties to manage. One of the products, for which delivery is usually outsourced, is a wedding cake. Ordering a wedding cake illustrates a typical Upstream process:

1. General research (recommendations from friends, family, and the most popular web pages or ranking platforms).
2. Narrow down options to those fitting most of the initial criteria.
3. Contact vendors and discuss the requirements to find the best fit (including delivery date, which is the wedding date).
4. Commit to one vendor and place an order.
5. Receive delivery, enjoy your day.

The wedding cake is an item with a known, fixed date, and delivering it after the wedding day is not a viable option. We know the delivery date, we know the risk (it's a Fixed Date item; missing the date spoils the wedding), and we can try to hedge the risk:

- Do proper research.
- Choose only an experienced bakery with a good reputation, ideally specializing in crafting wedding cakes.
- If you are very risk-averse, order a second cake from another place. The backup will be there, and the guests will probably eat an extra cake anyway.
- Or order the cake far in advance, such that you take delivery early. If it doesn't arrive, you have time to switch to a backup plan—perhaps a simpler cake from a different vendor. The trade-off is that the cake must be stored for a longer time and will not be so fresh on the day of the wedding. This requirement constrains the type of cake and the suitable recipes.

The wedding cake problem illustrates only one situation; when we encounter a cost of delay in daily life, we can handle it intuitively. Let's try to apply this concept to Upstream work.

Classes of Service

Generically, a class of service is defined by a set of policies that determine how something should be treated.[65] With kanban systems and workflows, one key aspect of a class of service indicates how the item should flow, or the priority that it should be given as it is pulled across the kanban board.

Lower-maturity organizations are advised to stick with the canonical set of four classes of service that originally evolved during an early Kanban implementation in 2007 (at Corbis). These four classes are focused on urgency and priority of pull through the system: Expedite, Fixed Date, Standard, and Intangible. These are illustrated in Figure 48 and described in detail in the *Kanban Maturity Model* Appendices "Triage"[66] and "Cost of Delay."[67]

Typically, the policies for the Standard class of service involve first-in-first-out (FIFO) queuing for Standard class tickets, or a policy to always pull the oldest available Standard class ticket.

As a general rule, any class of service with policies other than FIFO queuing negatively affects flow and adds delivery risk to a Kanban system. When multiple classes of service are present, the tail of the lead time distribution is longer and fatter. Classes of service are used as a trade-off: We trade predictability for a focus on value or risk. Classes of service are used to mitigate or react to the business or other external risks associated with a work item. The more urgent, critical, or valuable an item, the higher its class of service is likely to be.

More mature organizations may find it necessary to define their custom classes of service based on a detailed understanding of risks and trade-offs.

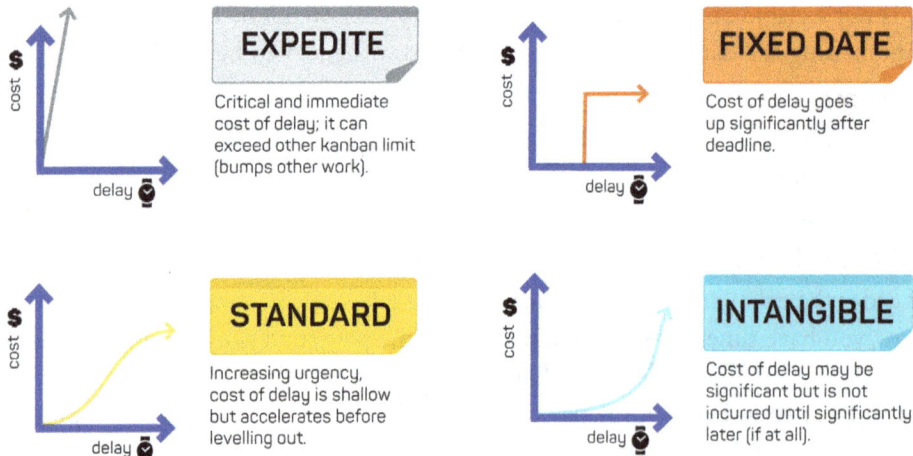

Figure 48 Four archetypes of classes of service as depicted in the *Kanban Maturity Model*

[65] D. J Anderson and T. Bozheva (2021), p. 194.
[66] D. J Anderson and T. Bozheva (2021), p. 409.
[67] D. J Anderson and T. Bozheva (2021), p. 403.

Note: The terminology used in this book when talking about the delivery dates is as follows:[68]

Generally, we use the term *deadline* to imply an artificially imposed date, by a customer, as a control mechanism—the customer doesn't trust the service delivery and is using the deadline as a forcing function. This is common at Maturity Level 2.

A *fixed date* is used to reflect an externality such as a public holiday, a regulatory date, or some other event that is normally beyond the control of either the service delivery organization or the customer. A fixed date is a risk that the customer is managing, whereas a *deadline* represents the customer shifting the risk to the delivery organization.

As trust improves, it should be possible to replace deadlines with service level expectations or, in a more formal, bureaucratic, and conservative culture, a service level agreement (SLA, a contract). In this case, the due date is defined by the replenishment commitment point plus the expectation of the service level. Due dates are now about ensuring trustworthy, predictable service delivery. They represent a benchmark, or target, for the healthy operation of the service.

For example, a wedding date is a fixed date, not a deadline. While it has been chosen and the choice was under the control of the couple getting married, it is not an artificially imposed date or arbitrarily set. A deadline would be a shotgun wedding where the father of the (pregnant) bride is pushing for an early wedding before the baby is born.

The four classes of service named by default are as follows:

- **Expedite**: There is an immediate, steeply rising (opportunity) cost of any delay.
- **Fixed Date**: At a date in the relatively near future, within twice the range of the lead time distribution, we expect a step function impact (or a sloping S-curve function very close to a step function) on a known and very specific day. This is typical of seasonal demand or specific named events, such as sports tournaments or public holidays.
- **Standard**: The cost of delay rises over a period defined by the range of the lead time and the range of the life cycle, initially in a convex pattern that eventually becomes concave. The shape is an S-curve over a medium to long time.
- **Intangible**: As the name suggests, the cost of delay is intangible or insignificant for a period longer than one to two cycles of the lead time range. While we model the Intangible class of service as having a convex function shape, we learned that these are, in fact, also S-curve functions. Intangible class of service work is commonly observed as platform upgrades; incorporating new components or technologies into an existing design; or redesigning, reworking, or generally maintaining the system for a product or service or its design or source code.

With respect purely to the CoD risk, we now know that this set of four classes of service provides sufficient coverage to adequately manage the temporal risks related to delay and its impact over time. It is no surprise, then, that this set of four classes of service has proven robust and is in broad usage in Kanban implementations globally.

---

[68] D. J Anderson and T. Bozheva (2021), p. 419.

Implementation Guidance[69]

- Visualize the classes of service:[70]
  - Assign a color scheme for the defined classes of service.
  - Create standard work policies for the use of colors in kanban boards across the entire business unit or organization. This enables managers and external stakeholders to correctly interpret the meaning of colors across all boards within the same product line or business unit.
  - Alternatively, use a horizontal lane on the board for each class of service, labeling the lane with each class's name.
- Use the class of service information to decide which work item to start next as well as which work items to select for the input queue at the Replenishment Meeting.
- Establish limits on the number of work items per class of service to ensure a balanced system, avoiding trends to processing mainly urgent work items.
- Use the lower class of service to compensate for variability in the number of work items in the higher classes of service.

Supporting Cases

Case 1: Posit Science[71]

Among other improvements, Posit Science wanted to address dysfunctional planning and prioritization. Janice Linden-Reed, who worked as a project manager at Posit Science, was looking to bring some organization and collegial collaboration to the process of selecting, sequencing, and scheduling work. To do that, she introduced classes of service to the Replenishment Meeting.

The concept was simple: Ask business owners to describe the impact over time for a given function. This would enable a determination of urgency. The discussion would help facilitate scheduling as well as the class of service required after the work was selected and committed. Stakeholders were briefed on the concept and asked to select the delay cost function that best matched the business risks associated with the request. This worked incredibly well. It was perhaps the most easily adopted of any new technique Janice introduced at Posit. It was quickly institutionalized, and years later was still in use for assessing risk and selecting and scheduling work. Their Replenishment Meeting board is shown in Figure 49.

---

[69] D. J Anderson and T. Bozheva (2021), p. 194.
https://all.kanban.plus/en/content/kanbanplus/kmm/default/kmm-posters/poster/kmm-metrics-poster
[70] D. J Anderson and T. Bozheva (2021), p. 144.
[71] Read more about Posit Science at: https://all.kanban.plus/en/content/kanbanplus/bwk/default/bwk-case-studies/reader/posit-science

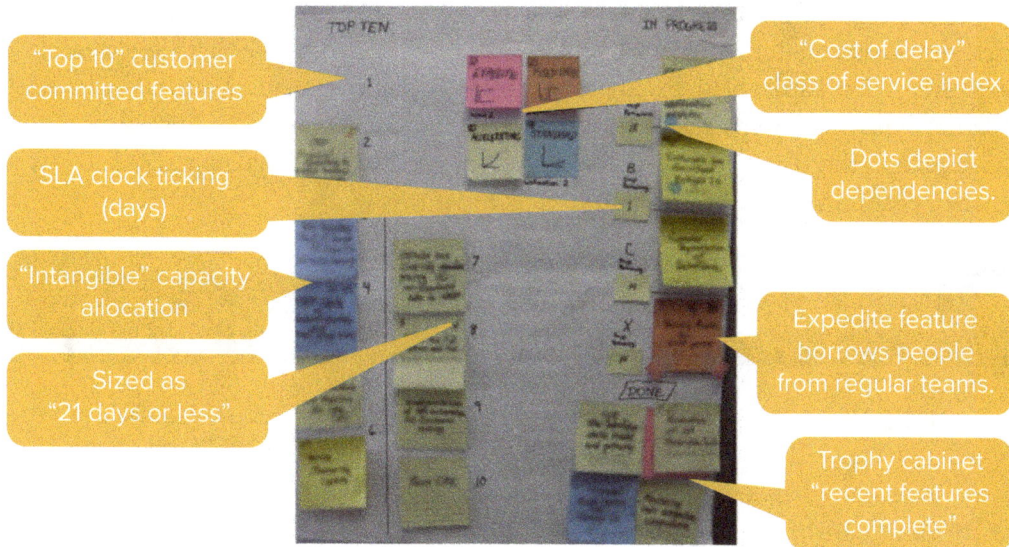

"Top 10" customer committed features

SLA clock ticking (days)

"Intangible" capacity allocation

Sized as "21 days or less"

"Cost of delay" class of service index

Dots depict dependencies.

Expedite feature borrows people from regular teams.

Trophy cabinet "recent features complete"

Figure 49 Posit Science Replenishment Meeting board

The *Top Ten* shows the input queue. In this case, it is a queue numbered one through ten. However, work wasn't necessarily pulled from the queue in a strict, prioritized order. When a team finished a feature and was ready to pull another, the first item in the queue might not have been a good match for them. They would work their way down the queue until they found the first strong match for their skills.

The Top Ten has a capacity allocation of two slots for the Intangible class of service work.

The legend shows the classes of service, the colors of the tickets, the delay cost function sketches associated with each, and any capacity allocation or other policies related to the classes of service. In this case, Expedite items are limited to one, and the Intangibles have a minimum of two.

Case 2: Tupalo[72]

At Tupalo, the classes of service were introduced to help developers distinguish items of higher value or urgency from those for which they need not pressure themselves. The reality was that not every request was as urgent as the developers used to assume. They introduced four different colors of tickets:

- Yellow for *Standard* items
- Red for *Fixed Date*, high-priority items
- Blue for *Intangible* work
- Green for *Expedite*, urgent items

---

[72] Read more about Tupalo at: https://all.kanban.plus/en/content/kanbanplus/bwk/default/bwk-case-studies/reader/tupalo

Each ticket had a date showing when it was started. Due dates existed only on the red tickets, which were for requests from external partners or for those that were associated with events that Tupalo held. Examples of these were Summer and Christmas Community Parties, Superuser Events at special and unique locations for the most active users, Cash Mobs supporting shopping in small, local shops, and Tupalo Challenges. Tupalo's first kanban board is shown in Figure 50.

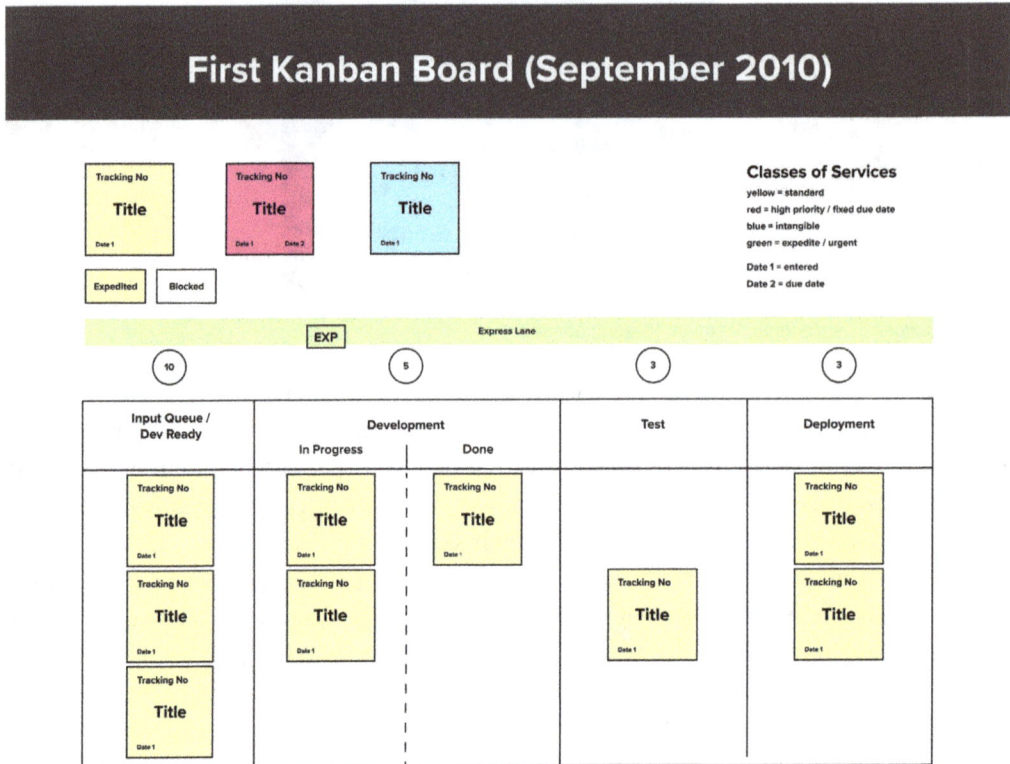

Figure 50 Tupalo's first kanban board shows the use of different classes of service.

Case 3: Department of Homeland Security

At the Department of Homeland Security,[73] triage for different work items (like defects or incidents) was visualized and managed using classes of service, as shown in Figure 51. The swim lanes on the board modeled the issues' severity.

---

[73] https://www.youtube.com/watch?v=M3H0AzmET7o

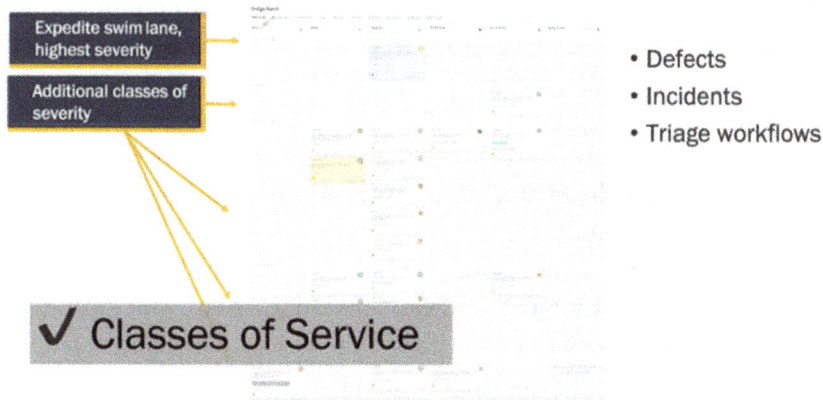

Figure 51 Visualization of classes of service at the DHS

The highest severity was shown at the top swim lane and the additional ones were shown in other swim lanes below. They didn't force the names of the four known archetypes; rather, they started with what they already had implemented and improved it using the logic behind the canonical classes of service.

Two-Phase Commitment Approach

The concept of a two-phase commitment[74] is only partially related to Upstream Kanban. The first commitment takes place at the Replenishment Meeting, which is Upstream. The second happens later, Downstream, during the Delivery Planning Meeting and relates to the commitment of a delivery date based on a level of predictability.

- The first commit phase happens when an item is selected and pulled into the kanban system at the Replenishment Meeting—the commitment to do the work.
- The second commit phase happens later, once the ticket is sufficiently far along in the workflow that its arrival in the ready-for-delivery buffer can be predicted with a high confidence level. This enables a commitment to a specific delivery date.

Mature Kanban implementations should separate the commitment to do the work from the commitment to deliver on a specific day. This better manages the psychology of the customer: They get a high level of confidence that the work is started and that it will be delivered—a promise you can keep—at the first phase of commitment; later, they get a precise, specific delivery date, one for which you have high confidence that you can hit and hence, keep the promise you are making.

Using two- or multi-phase commitments is a means to manage customer expectations. It is popular with, for example, furniture retailers, which rarely keep much stock on hand

---

[74] D. J Anderson and T. Bozheva (2021), p. 196.

and need to order items from a warehouse or manufacturer. They often promise only approximate delivery dates, such as ten to thirteen weeks, and then, as the delivery date gets closer and more certain, they add precision to their forecast and their promise—your dining table will be delivered in week thirty-one. Later, they'll say that your dining table is scheduled for delivery on Tuesday of week thirty-one. Finally, perhaps a day or two before that, you'll learn that the delivery truck will arrive with your dining table between eleven a.m. and one p.m. on Tuesday.

The multi-phase commitment has been well established in physical-goods industries to manage the psychology of the customer and their expectations. Start with an accurate but imprecise promise and slowly move to a more precise promise as uncertainty is reduced or eliminated.

Implementation Guidance

- The first phase of commitment happens at the Replenishment Meeting (see more in Chapter 4, Upstream Feedback Mechanism: Meetings and Reviews).
- Communicate that replenishment commitment is a commitment to get started and to deliver at some point in the future—a commitment to do the work.
- The second commitment phase happens during the Delivery Planning Meeting.
- The second commitment phase commits to a specific delivery date, so now a specific piece of work is scheduled as part of a specific delivery.
- After the second phase of commitment, items should be tagged with the target delivery date and have their class of service bumped up to Fixed Date to ensure that everyone involved understands that a specific date commitment has been made.

Supporting Case: Investment Bank

The work on the project in investment bank was very client-decision dependent, where the Business team had the final word on what and in what order work should be developed. They were involved in the Upstream process of collecting information, formulating requirements, and reviewing the design.

See more about the design elements in Chapter 2, Designing Upstream Kanban.

The two-phase commitment was successfully introduced as the answer to the commonly recurring question from the project stakeholders: "When can we start using what you are developing?"

The previous approach—namely, putting best-guessed, gut feeling–based dates into a Gantt chart—was a disaster. The commitments were never met, clients were angry and stopped believing any promises they were given, and each biweekly Steering Committee meeting was a theater in which everyone was just playing their roles, but there were no meaningful outcomes.

Something had to change, and among other changes, a new approach to communicating dates was introduced:

- **First Commitment Point—Replenishment Meeting:** When a new item was ready to be pulled (requirements ready and reviewed by the Business team), only the lead time was communicated to the Steering Committee. The information was dependent on the work item type, as the lead time was measured at this level. For example, if a new pulled item was a new report, the expected lead time for this type of work (85th percentile) was seventy-five days. Any potential deviations from expected delivery were reported to the Steering Committee on a biweekly basis.

- **Second Commitment Point—Internal Acceptance Testing:** When most of the test cases were completed and all major bugs had been resolved, the Project Team started preparation for the deployment. As it involved a few other company-wide teams to support the release, the deployment calendar was always known in advance for the whole year (including specific dates and time windows on the deployment afternoons). The Project Team had to fit the release into the nearest deployment window. Also, at this point, the specific deployment date was communicated to the Steering Committee, Business teams, and all other stakeholders.

Figure 52 visualizes the two-phase commitment approach at investment bank.

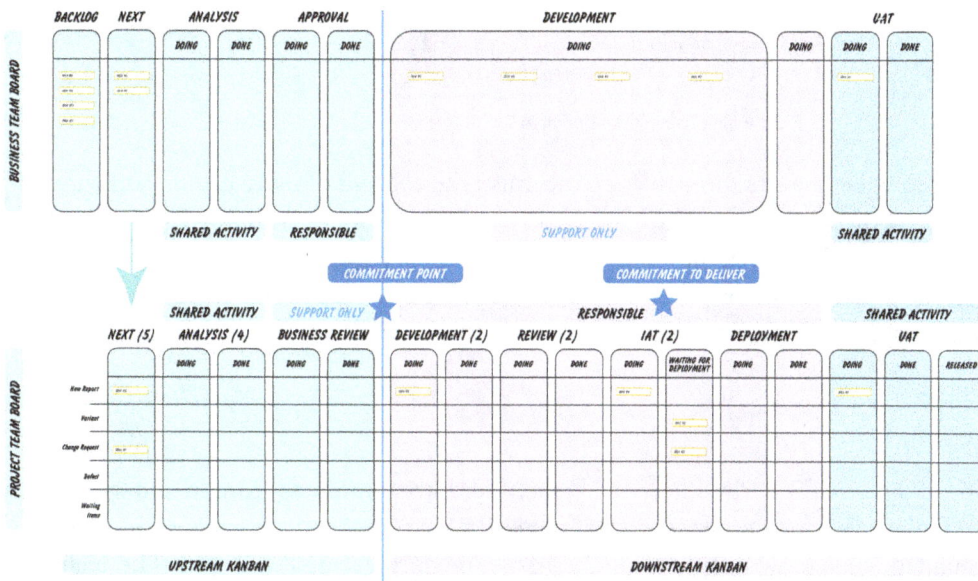

Figure 52 Two-phase commitment at investment bank

Supporting Case: FARA

The case of FARA development teams shows a lower-level implementation with the emerging process in which initial ambiguity was replaced by an asynchronous commitment to "what we deliver," but with no lead time information. In that sense, it was more like a triage decision on "what we do next" supported by the *Definition of Ready* mutually agreed by the Product team and the Development teams. This is illustrated in Figure 53.

See more about the design elements in Chapter 2, Designing Upstream Kanban.

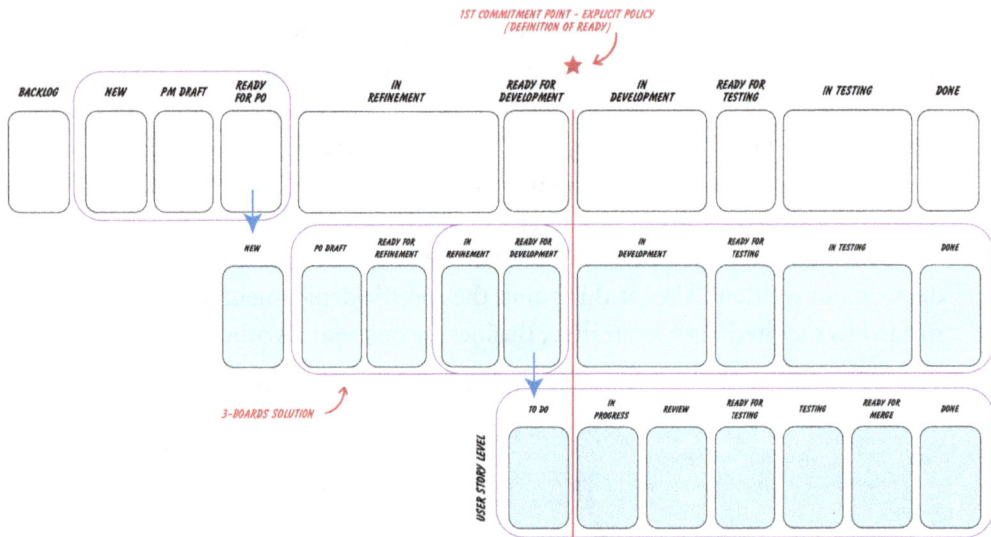

Figure 53 Visualization of the first commitment point at FARA

We'd been in a situation of low to no trust; the concept of mutual commitment didn't exist, and the work was pushed to delivery teams without considering their constraints. Hence, establishing even the most basic "Definition of Ready" policy sparked communication and collaboration between the Product and Delivery teams and helped create a better understanding of what we wanted to—and actually could—deliver.

Triage

A Short History of Triage

The term "triage" is derived from the French word "trier" meaning "to sort" and was initially used for sorting food products such as coffee. In the early 1800s, the approach of "sorting" wounded soldiers was applied for the first time during Napoleon's war. The technique developed during the Crimean War (mid-19th century) when the casualties were divided into four groups in order of care priority:

- The mortally wounded were assigned to the care of the Sisters of Mercy (hospitals).

- The seriously wounded who required urgent surgery received it at the emergency dressing station.
- The less seriously wounded were transferred for surgery the next day.
- Those who had minor wounds were treated and returned to their units.

Today the term "triage" is routinely applied to both military and civilian situations in which the number of casualties is expected to exceed the care facilities available in the area.[75]

There are two elements embedded in the concept of triage from the start:

- Constrained facilities and/or limited capacity of medical personnel.
- Differentiated treatment of people in the system based on specified categories, procedures, and rules.

Both elements also are present when applying this technique to knowledge work.

Triage in Knowledge Management Work

Triage is a technique that works for managing the queuing of work for any scarce resource or service with limited capacity. As soon as we have a WIP-limited pull system, a Kanban system, we need to triage. The work in the system is the work we are doing now, while work queuing Upstream is work for later; and if we discard an option, that is work that we decided not to do at all. Upstream Kanban, or analysis required for a definition of ready, or the discussions at a Replenishment Meeting can all be used to implement an effective triage discipline.[76]

Triage is intended to avoid overburdening a system and provide rudimentary prioritization, or selection and sequencing criteria. The traditional approach to triage is that the work should be separated into three basic categories: now, later, and never.

- Now! (It's likely to "survive" only when given immediate treatment.)
- Later (It's not urgent; it can wait.)
- Not at all:
  - It's not likely to "survive," no matter what we do.
  - It will be fine, no matter what we do, or can be solved in another way, or by another (service) provider (the equivalent of "go home, take two aspirin, you'll be fine in the morning").

Items for later may receive some further categorization. In those cases, it is important to establish "approximately when." Setting these expectations has the psychological benefit of alleviating anxiety over the belief that "if not now, then never."[77]

---

[75] https://www.civilwarmed.org/surgeons-call/triage/

[76] D. J Anderson and T. Bozheva (2021), p. 403.

[77] D. J Anderson and T. Bozheva (2021), p. 195.

"When" relates also to the changing pattern in cost of delay, and when this function (or curve) is known, even approximately, it informs the "when." For example, a patient bleeding slowly will not die soon, but they will eventually bleed out and die if not attended to later. The rate of blood loss determines the time window and the "when." For work items, the time or date can be later, while it will be (mutually) understood that it is not too late—it is not beyond the last responsible moment to act.

Over years of applying triage to the decision-making process, we soon realized that triage is not one but two decisions![78]

> Further examining and elaborating on Upstream processes and options development revealed that when there was just a backlog, or a big pool of options, and a Replenishment Meeting, traditional triage seemed like the answer. But with detailed Upstream, and embedded option points for invest, shelve, and discard, it exposed that the decision of what is "important" (and the decision of Should we do it or not?) is separated from the cost of delay.
>
> It also enabled a simpler solution in that the cost of delay can be considered relatively for any one item on its own, rather than attempting to quantify it for comparison against other available options. This also simplified the problem that apples cannot be directly compared with oranges. For example, table-stakes features—must-have items—have a binary impact: Either we include them and the product is viable or we don't and it isn't. However, they have little intrinsic value beyond that—people will not say, "I bought the car because it had a steering wheel"—so no value can be attributed to it.

The two triage decisions are:
- **Decision #1: Should we do it or not?**
  - We need a mechanism to understand whether what we plan to do is valuable enough.[79]
  - Decision is made in the Upstream.
- **Decision #2: When should we do it?**
  - If we decide that something needs to be done, when should we do it?
  - We need to understand when is "now" versus the desired delivery date.
  - It may require both Upstream and Downstream (people) to make the decision, as it requires an understanding of lead time and delivery and schedule risks.

These two decisions are not made at the same time, and they require two different skill sets of different roles in the organization.

---

[78] The original observation is attributed to Alexei Zheglov.

[79] The value is assigned by our strategy, our target market segmentation, and the fitness criteria for that market. The danger is always to think of "value" as "money." Something may be valuable because it aligns with our brand essence and company values, but it doesn't directly make us any money.

> **Note:** Two decisions are possible only at deeper maturity levels where refutable demand is recognized. It requires:
> - a more elaborate Upstream with embedded options,
> - a planning/commitment meeting, separate from a replenishment/pull meeting, or, alternatively,
> - the same meeting but with different cadences so that every second, third, and fourth Replenishment Meeting is extended to include a planning/commitment activity, with a subset of the attendees.

At lower maturity levels, where all demand is irrefutable, the first decision (Should we do it or not?) is already made, and commitment has happened. Hence, only *when* must be decided, which can be as simple as *now* or *later* (in other words, select or defer; there is no discard option). Then, the follow-up decision will be made at the next Replenishment Meeting.

Implementation Guidance

- Triage is most commonly associated with prioritizing defects for fixing. It can be equally applied to new work requests and managing blocking issues.
- Establish explicit policies and criteria for answering the triage questions:
  - Should we do it or not?
  - When should we do it?
- Decide who is responsible and accountable for making these decisions.

Start Date Ranges

Figure 54 shows a visualization of the start date ranges[80] relative to the desired delivery date.

To understand and use the ranges, you must first know the desired delivery date (DDD) of the work item and its lead time.

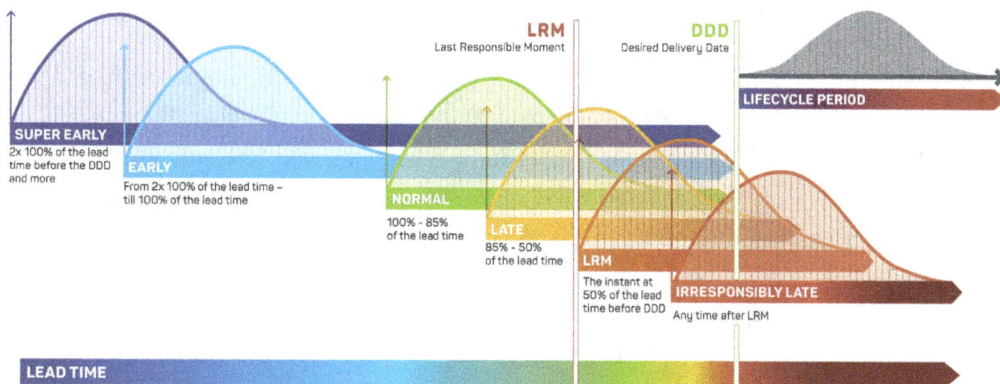

Figure 54 Start date ranges as depicted in the *Kanban Maturity Model* (Triage Tables)

---

[80] D. J Anderson and T. Bozheva (2021), p. 411.

DDD is the ideal date to take delivery. For example, a cost-saving feature can have an impact tomorrow if we receive it today, and hence, the DDD is today. However, for many items that are seasonal or tied to specific windows of opportunity for a product or service, the DDD is in the future. A marketing campaign for Easter-themed items would not be launched during the year-end holiday season. The DDD is likely to be sometime in January; for example, the third Tuesday in January, because we always launch campaigns on Tuesday, and we want about ninety days of life from our Easter campaign.

- **Super Early** is more than two times the 98th percentile of the lead time distribution range before the DDD, or before the item is needed.

  This choice is so far ahead of when we need the item that it could be a background task, frequently delayed to prioritize more urgent work, or we could afford time to throw our work away twice, do it over, and yet still deliver within expectations without incurring any cost of delay.

- **Early** is more than one full lead time range before the DDD.

  This choice means that we could afford to throw away our work and implement a do-over and probably still deliver on time without incurring any cost of delay.

- The **Normal** range lies in the tail—the 100th percentile down to the 85th percentile of the lead time distribution range before the DDD.

  This choice gives us a very high probability—at least six out of seven—that we will deliver on time without incurring any cost of delay.

  Beyond this, we consider that we are starting later than would be ideal.

- The **Late** range is from the 85th percentile of the lead time distribution to the 50th percentile (or median). We consider the median, where there is only a one in two chance of on-time delivery, to be the last responsible moment (LRM). This gives mathematical meaning to a term that has been in common usage, without any rigorous definition, in Lean literature for twenty to thirty years.

  In this range, we can have a coin-toss chance of incurring some cost of delay. Management attention needs to be paid to the item to keep it flowing.

- Beyond the LRM, we consider that an item is being started **Irresponsibly Late**.

  Our chance of on-time delivery is worse than the chance of a coin toss landing in our favor. Without heroic effort and expediting work, the item must incur some cost of delay.

  The Late and Irresponsibly Late start times have the impact of stressing the organization and consuming management time and attention that would be better spent on more strategic, system-level activities.

Start date ranges provide you with additional information that supports the decision-making process in terms of the urgency of work and the proper treatment of the work items. Those that are Super Early probably do not require special attention and do not create urgency. Those that are Late may need to be expedited. Those that are Irresponsibly Late may require asking, "Has it already bled to death?"

Refutable and Irrefutable Demand

There are circumstances in which the demand for an internal shared service is likely to be irrefutable. These requests happen because someone accepted a customer request without verifying the delivery capability. For example, the Sales team is pushing for new contracts disregarding the delivery team's full capacity. In such a situation, discarding an option is not available, because the work has never been optional—it was committed from the start. The choices are limited to shelve or invest.

That requires differentiating when an organization has a formal Upstream versus those that have just a request pool.

With refutable demand, we can afford two triage decisions: Should we do it or not? and When should we do it?

When demand is irrefutable, the commitment has already happened, and the first decision (Should we do it or not?) has already been made, so we need only to decide when we should do it. Our options are limited to shelve or invest immediately, so we can either select or defer. We cannot discard.

Even though the demand coming from customers is irrefutable, and the first triage decision was made outside or beyond our control, there still exists some degree of optionality. It is, however, limited to **internal optionality**, the decision about **"how we do things"** or **"how we build the solution."** We may explore a couple of options for potential solutions, and that's where all three options (invest/shelve/discard) still exist—but internally only.

We need to develop the skill to differentiate the truly irrefutable demand, such as:

- Legal or regulatory requirements
- Table stakes for the customer or market niche
- A mission-critical feature
- High-severity production defects
- A necessary piece—part of a bigger whole, without which the larger item is unfit for purpose and will fail to meet expectations

Demand that is artificially irrefutable and dysfunctional—such as the boss wants it or we've already committed to it—requires negotiations and establishing a clear and explicit policy.

Implementation Guidance[81]

- Irrefutable demand can still be scheduled, enabling evenness of flow.
- Negotiate deferred commitment with customers.

---

[81] https://all.kanban.plus/en/content/kanbanplus/kmm/default/kmm-content-overview/reader/kmm-module8-videos

- Use the feedback loops of Risk Review and Operations Review to balance demand with capability.
- Introduce classes of service based on risk.
- If overburdened with demand, bifurcate and route low-risk items to where there is slack capacity—effectively trading non-functional quality or fidelity for shorter lead time.
- At a customer-facing point, insert a strong "definition of ready" that states that we will commit to external demand only when internal dependency capacity is confirmed.

Planning

Now, how can we use all these pieces of knowledge to answer the introductory question from the beginning of this chapter: "When should we start to deliver on time?"

Let's put the puzzle pieces together. To do that, we use a technique I used to call "the planning recipe."

The recipe has four data-driven ingredients:

- What is the lead time?
- When is "now" (or the expected start date)?
- What is the desired delivery date?
- What is the time frame between "now" (or the start date) and the desired delivery date?

And one additional, but essential, ingredient:

- What is the customer's expectation? For example:
  - What is the cost of delay? And how does its rate change over time? What is the shape of the cost of delay curve?
  - Should we change the class of service?
  - Can it wait?

We combine this recipe with triage and a two-phase commitment approach in the following example.

Supporting Case: Investment Bank

In this example, the project was to replace a set of manually prepared reports with automatic reporting. The project team consisted of Business Analysts, Developers, and a Project Manager, with the support of multiple shared services from the organization.

The decision-making process related to developing new automated reports and their variants was owned by the Business team (an internal client) and their representatives called the Ideation Group. Figure 55 visualizes the process using the boards of both the Business team and the Project team (see more about the design elements in Chapter 2, Designing Upstream Kanban).

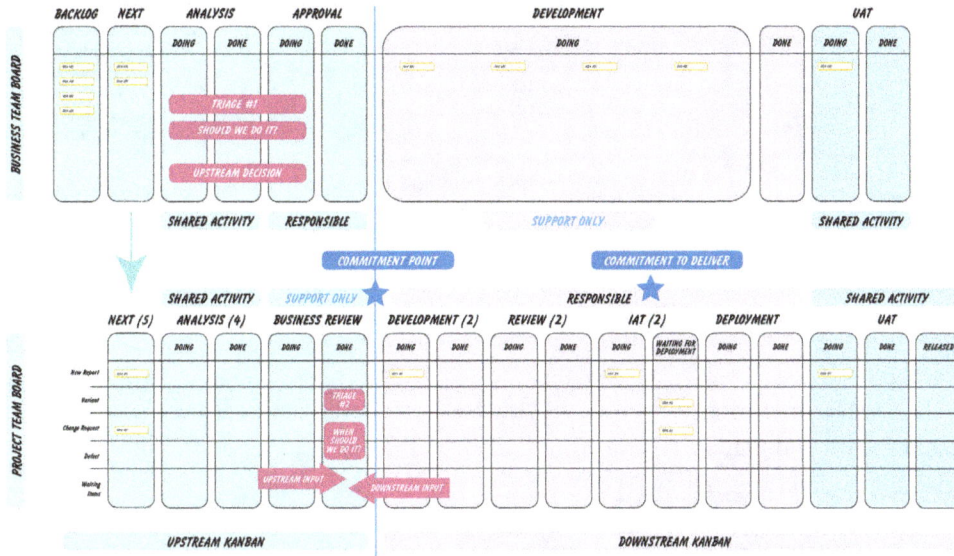

Figure 55 Visualizing the process using the boards of the Business team (Ideation Group) and the Project team

New items were pulled from the *Backlog* column to *Next*, and then to *Analysis*, where project Business Analysts, together with respective business Subject Matter Experts (SMEs) from the Ideation Group, collected information, put it into the business requirements (including risk analysis and regulatory requirements), and drafted the initial shape of the report for the developers to create.

Part of this process was answering the first triage question: Should we do it or not? The main criterion for deciding whether a report was eventually moved to *Business Review* (reviewing and approving requirements by the Team manager, who was normally the most experienced employee with the authority to decide) was balancing off the effort required to automate the reporting with the actual time savings on the Business side. The time-saving aspect consisted of elements like the frequency of preparing the report, the number of Financial Analysts needed to do it, and the number of external clients affected.

The final decision on whether we should do it was made by the Team manager in the process of reviewing and approving requirements.

Whenever a report was accepted and moved to the *Business Review Done* column, the second triage question came into play: When should we do it? Or, more precisely in this case: Which one should we start next and when?

This is where the planning recipe comes in. Let's look at the questions one more time, starting with data-driven information. To support answering the questions, we use a calendar visualization, as in Figure 56.

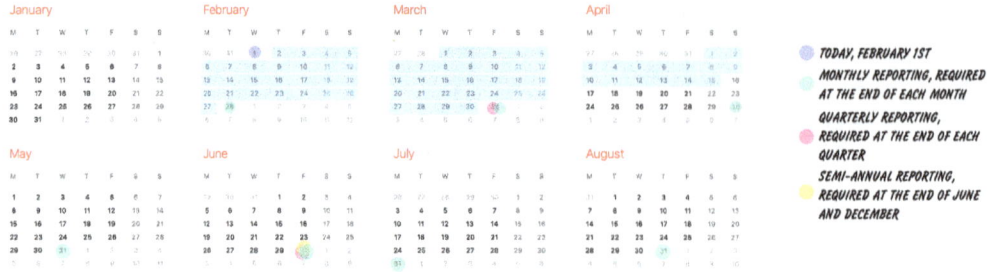

Figure 56 Using a calendar to support the decision-making process on the second triage question: When should we do it?

In our case, three reports for monthly, quarterly, and yearly periods were accepted by the Business team and manager as meeting the criteria for Should we do it? These are shown as the green, pink, and yellow reports.

Question #1: What is the lead time?

Based on our historical data, we knew that the lead time (at the 85th percentile) for *new* reports was seventy-three days, forty days for the *variants* of the reports, thirty-five days for *change requests*, and five or twenty-five days for the *defects* (depending on the deployment calendar).

For simplification in this supporting case, let's focus on the *new* reports, so the answer to Question #1 is seventy-three days (it is visualized as the blue area spanning from February 2 to April 15).

Question #2: When is "now" (or the expected start date)?

In our case, "now" is February 1.

Question #3: What is the desired delivery date?

The desired delivery date depends on the type of reporting. This could be either monthly, quarterly, or semi-annually. The calendar presents the reporting dates falling between February 1 and August 31.

Question #4: What is the time frame between "now" (or the start date) and the desired delivery date?

As above, this one also depends on the reporting period:

- The most immediate report that could be delivered was thirty days away. It was, however, a monthly report (green), so the delay was always thirty to thirty-one days. The start date range for February was *Irresponsibly Late;* for March, *Late;* and for April, *Early* (close to *Normal*).

- The next, a pink report, came due on March 31. The next opportunity to realize for automating it (if the March date was missed) was at the end of June. The start date range for March was *Late*, but for June it was *Early* (close to *Super Early*).
- The last report, the yellow one, was due at the end of June. The start date range in this case was still *Early* (close to *Super Early*).

Considering **only** those elements of the recipe that were provided by the Project team, and if it was up to **only** the Project team to decide, they would choose a green report with the end-of-April delivery date.

But there is a reason why the recipe has the fifth ingredient, and why we say that the second triage decision may require (and usually requires) both Upstream and Downstream input.

Question #5: What is the customer expectation?

What did the Business say about the choice that we had?

- The green report is a monthly task for one client. It consists of one small spreadsheet. The cost of delay is intangible!
- The pink report is required by 80 percent of 120 clients. Automation will save dozens of work hours. The cost of delay is very high!

To properly manage this situation, the class of service of the pink report was changed—the report was promoted in scheduling—to complete it for the next reporting period (end of March). To ensure the smoothest possible flow of work, the Business allocated additional SMEs to support the process of testing and answering developers' questions. The Project team cleaned up the schedule from less urgent items that could be delayed with low to no impact.

At the point of mutual agreement (synchronous commitment), the work item was pulled from *Business Review Done* to *In Development*, and the work in Downstream started. That was considered the first commitment point.

This is where involving both Upstream and Downstream in answering the second triage question is beneficial:

- Using only data without the customer's input, we wouldn't have a full picture, as the cost of delay and urgency component was missing.
- On the other hand, using only the cost of delay and urgency determined by the customer causes the delivery teams to be overloaded and not able to finish anything on time, as the decision-making process is detached from the data.
- Incorporating both parts in the discussion ensures that the proper work is started on time and the delivery part of the transaction is not flooded by unlimited customer demand.

Takeaways

- A good approach to work planning is to replace "When will it be done?" with "When should we start so that we deliver on time?"
- The cost of delay has two aspects: the actual cost and the opportunity cost.
- A class of service is defined by a set of policies that determine how something should be treated.
- Mature Kanban implementations should separate the commitment to do the work from the commitment to deliver on a specific day.
- The term "triage" is derived from the French word "trier," meaning "to sort," and was initially used for sorting food products such as coffee. In the early 1800s, the approach of "sorting" wounded soldiers was applied for the first time during Napoleon's war.
- Triage is a technique that works for managing the queuing of work for any scarce resource or service with limited capacity.
- Today, we see triage as two decisions: Should we do it or not? and When should we do it?
- There are circumstances in which the demand for an internal shared service is likely to be irrefutable. The choices are limited to shelve or invest.
- We need to develop the skill to differentiate truly irrefutable demand from demand that is artificially irrefutable and dysfunctional.

Closing Thoughts

This book focuses specifically on Upstream Kanban. I made some deliberate omissions to avoid repeating content from other Kanban books you may already know.

- If you are interested in learning what the Kanban Maturity Model is, check Chapter 13 of *Discovering Kanban*,[82] the first book in the Better with Kanban series.
- When you find that useful and applicable to your business context, then I suggest going more deeply and studying the entire *Kanban Maturity Model*.[83]
- Most of the case studies referred to here can be found in BWK+,[84] KMM+,[85] and on the Kanban University YouTube channel.[86]
- All practices that will help you with designing the Downstream Kanban system will be described in depth in the forthcoming second installment of the Better with Kanban series, *Implementing Kanban*.
- To learn more about designing and building products and services that are fit for the customer's purpose, I truly recommend the book *Fit for Purpose*.[87] It will help you answer questions such as "How do I know who my customer is?" "How do I know that the improvement I introduced brings real value to the customer?" "How can I learn what my customers really need?" "Do I measure the right things?" and so on.

---

[82] David J Anderson, *Discovering Kanban* (Chicago: Blue Hole Press, 2023).

[83] D. J Anderson and T. Bozheva (2021).

[84] D. J Anderson (2023).

[85] D. J Anderson and T. Bozheva (2021).

[86] https://www.youtube.com/@KanbanUniversity

[87] D. J Anderson and A. Zheglov (2023).

"Pragmatic, actionable, and evidence-based guidance" is the motto of each of our classes!

All the practices and implementation guidance described here were tested and introduced in real Kanban implementations around the world. Those that were presented by the case owners during conferences are directly referred to in the text. There are also multiple other, smaller cases of introducing Upstream Kanban practices shared by students and trainers over the four years of delivering the *Kanban for Design and Innovation* class.

I'm certain that there are many more Upstream Kanban implementations around the world! Kudos to everyone for doing it! Please share your experiences with me; we will make sure that the global community learns about your achievements!

Thank you for your time and interest in Upstream Kanban and all the best on your journey!

Appendix
Understanding Lead Time[1]

U nderstanding lead time is necessary as a means to break out from Maturity Level 1, Team-Focused, and move to Maturity Level 2, Customer-Driven, and then beyond to Maturity Level 3, Fit-for-Purpose. Lead time isn't a single number; rather, it is a probability distribution because the time something takes from commitment to delivery is always non-deterministic. Effectively managing a professional-services business requires a basic understanding of mathematics not typically included in the curriculum for secondary education. The nature of the observed probability density functions (PDFs) associated with lead time is also not commonly taught in undergraduate-level one- or two-credit classes in statistics. Hence, it is necessary to provide some mathematical background to underpin the application of lead time distribution functions to real-world problems such as forecasting and managing cost of delay risk.

Definition

We define lead time as starting at the mutually agreed commitment point and continuing until an item is ready for delivery. In other words, we start counting lead time from the point when a customer can legitimately expect us to work on an item until we can legitimately expect the customer to take delivery. Any additional waiting time on the delivery end is not counted. This definition is unambiguous and works consistently for any workflow or service.

You should, however, recognize that this definition works reliably at Maturity Level 3 and deeper, while at lower maturity levels it is problematic. At lower maturity levels, the commitment point is often ambiguous and at best asynchronous; that is, the customer or the upstream part of the workflow commits a request before the delivery or downstream service workflow is ready to commit. In this type of asynchronous commitment, we measure lead time from the second half of the commitment, the point when the delivery workflow pulls the item into WIP. In Maturity Level 2 workflows, often there isn't a strong concept of commitment, and hence, there is a tendency to measure lead time from the point a work item is submitted. This is legitimate for internal shared services and other similar systems where the work is irrefutable.

---

1. Reproduced from Appendix C in *Kanban Maturity Model*, 2nd Professional Coaches' edition, by David J Anderson and Teodora Bozheva (Seattle: Kanban University Press, 2021).

For clarity, and most relevant in a lower-maturity world or a maturing and slowly scaling-out Kanban implementation, we use the terms "customer lead time," as defined above, and "system lead time," as a degenerate form, describing the time from when an item is pulled until it is ready for delivery. Customer lead time (or just plain lead time in a higher-maturity implementation) is a fitness criterion, while system lead time can only ever be a health indicator or an improvement driver. The difference between customer lead time and system lead time is always a non-deterministic period of waiting on the front end, the time between when the customer submits a request—or believes it was committed—to when it is actually pulled into the kanban system and actually is committed. In a Maturity Level 3 or deeper kanban implementation with synchronous commitment, customer lead time equals system lead time: they are equivalent. The effect of a difference in customer versus system lead time is that the customer lead time suffers from a longer, fatter tail, and possibly also has a higher value for location, all of which is explained in the next section.

The term "lead time," therefore, can be explained this way: "If you know when you need to take delivery, the lead time is the amount of time in advance you need to place an order to expect delivery on or before when you need the item" or, simply, the period of time that commitment must precede delivery.

Nature of Lead Time

The inconvenient truth is that lead time is not a single number; rather, it is a random number defined by a probability distribution function.[2] The good news is that by recognizing the shape of the lead time distribution, we can learn a lot about the nature of the work and the delivery risks associated with it. This enables us to make effective risk-management and planning decisions that provide for a reasonably optimal economic outcome given the uncertainty involved.

Figure A.1 shows two histograms of lead time data from real kanban system implementations. The x axis shows the number of days of lead time, and the y axis shows the number of occurrences of that time within the sample data set (the number of items pulled through the kanban system). The one on the left is from an IT operations group and the one on the right is from a software product development team. The one on the left is said to be fat tailed: there is a long, visible tail stretching out to the right along the x axis. Generally speaking, fat tailed is undesirable, risky, and makes planning difficult. The one on the right is said to be thin tailed: there isn't a long, visible tail running off to the right along the x axis. This naming convention can seem counterintuitive to non-mathematicians. The truth is that mathematically, these tails run to infinity. Hence, if you can see the tail, it must be fat—a thin tail is effectively invisible along the x axis. These concepts of fat-tailed and thin-tailed lead time distributions are very important. Knowing which one you have in your kanban system makes

---

2. Technically, it's a probability density function.

a vital difference in planning, risk management, and the likelihood of achieving customer satisfaction and being viewed as a trustworthy service provider.

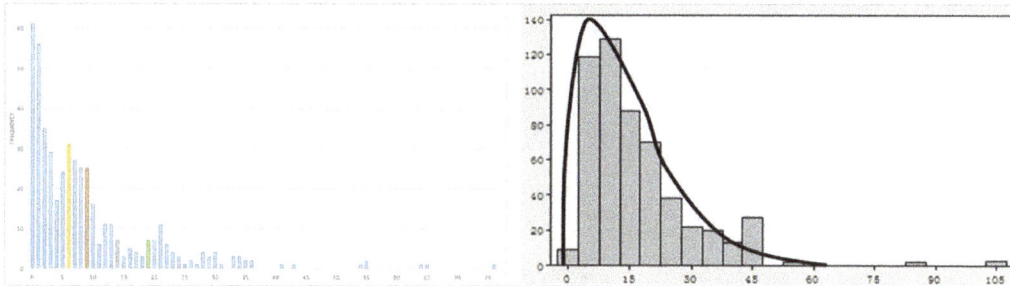

Figure A.1 Two examples of lead time distributions

Assuming you have single-modal data (see below), there are some simple ratios that indicate the likelihood that you have a thin-tailed or fat-tailed distribution:

$$\frac{(98\%ile - location)}{(50\%ile - location)} \; > = 5.6 \; => \; \textit{fat tailed}$$

$$\frac{(98\%ile - location)}{(50\%ile - location)} \; < 5.6 \; => \; \textit{thin tailed}$$

For convenience, you will sometimes see these written in a simplified form, assuming location = 0 and that the 98 percent is effectively the end of the tail you can see. Hence,

$$\frac{tail}{median} >= 5.6 \; => \; \textit{fat tailed}$$

$$\frac{tail}{median} < 5.6 \; => \; \textit{thin tailed}$$

These appear in the poster for Triage Tables available from Kanban University.
A further sanity check is available for thin-tailed distributions,

$$\frac{(98\%ile - location)}{mode - location} < 16 \; => \; \textit{thin tailed}$$

or simplified as

$$\frac{tail}{mode} < 16 \; => \; \textit{thin tailed}$$

This formula is not valid for fat-tailed distributions.

These results are produced through a thorough analysis of the behavior and attributes of the Weibull function, which is not included here.

Interpreting a Curve

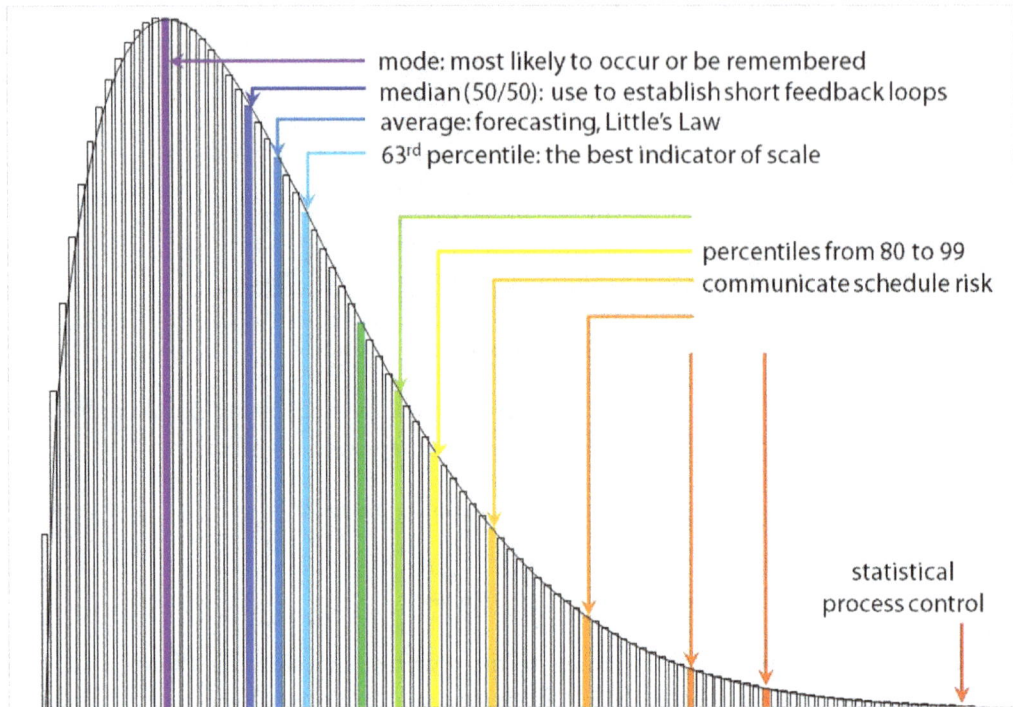

Figure A.2 Interpreting a lead time distribution curve

The mode is the top of the hill (purple in Figure A.2), the most commonly occurring lead time in the data set. The psychological impact is that it is the most memorable. If the mode is fifteen days, and you ask someone how long things typically take, they are likely to answer that it's usually around fifteen days because this is the most memorable, dominant result.

The median is the 50th percentile. If the median is twenty days, it means that half of the items processed took less than twenty days, while the other half took twenty or more days. This leads to the psychological result of "on average, things take twenty days" because half take less than twenty, the other half more. When humans use the term "average" to report a phenomenon they experience physically, emotionally, and with their sensory perception and limbic brain, they are almost certainly reporting their experience of the median.

The mean is the arithmetic average: sum up the value of all the data points and divide by the number of points. In plain language, the word "average" is usually used to refer to this arithmetic mean. Humans use this term correctly when they are referring to something they experience as mathematics and are processing with the logical brain, their prefrontal cortex. The mean tends to accelerate away from the mode and the median as the tail extends farther to the right with higher data points (work items with longer lead times). A fat tail affects the mean much more than it affects the median and is unlikely to impact the mode at all. Understanding this is important for planning, risk management, and customer satisfaction.

Planning is affected because simple forecasting equations, such as Little's Law and the use of regression to the mean, require using the mean. Just a few high-value data points skew the mean upward and may dramatically affect the accuracy of a forecast.

It is important to remember this simple mantra, "The risk is always in the tail." Fat-tailed distributions require a different approach to managing risk. The next section explains this in greater depth.

Fat tails affect customer satisfaction. The impact of a single long lead time, a single high-value data point, can destroy customer trust. For example, one item has a mode of ten, a median of twenty, and a mean of thirty; as a customer, I ask, "When will my request be ready?" and I am told, "We usually process items in fifteen to twenty days." I then wait 155 days—or ten times longer than I'd been told to expect. Burned by this one bad experience, I no longer trust the service delivery. Consequently, every future request I make will have a deadline attached to it and penalties for failure to deliver.

Figure A.2 also illustrates a set of percentile points as the tail stretches to the right; for example, the 85th percentile, or the value where 85 percent of the data points are to the left (and lower), while 15 percent are to the right and of higher value. This is illustrated again in Figure A.3. The curve in Figure A.2 is said to be right skewed. Again, this may seem counterintuitive, as the bulk of the distribution appears to be on the left.

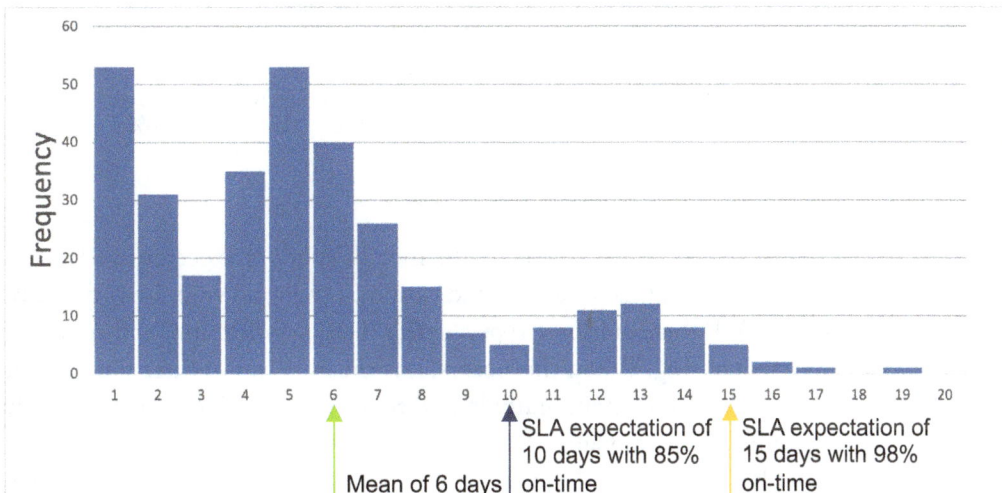

Figure A.3 Lead time distribution showing service level agreements

Figure A.3 shows a mean of six days, and the 85th percentile is ten days, indicating that six out of seven items are completed in ten days or less, while one in seven takes longer.

The 98th percentile of fifteen days indicates that only one item in fifty takes longer than this. We find that for most knowledge work—intangible goods industries—there is a psychological acceptance level of six out of seven. Therefore, the 85th percentile is a very powerful tool. Trust can be established by delivering six out of seven items within a promised period (e.g., ten days), providing the impact of missing the promised date is not severe. So, psychologically, I can trust a service with six out of seven on-time deliveries within a service level expectation (SLE), providing the tail is thin and the penalty for exceeding ten days is only a few days more. Thin-tailed lead time distributions are trustworthy; fat-tailed distributions are not.

Multi-Modal Data

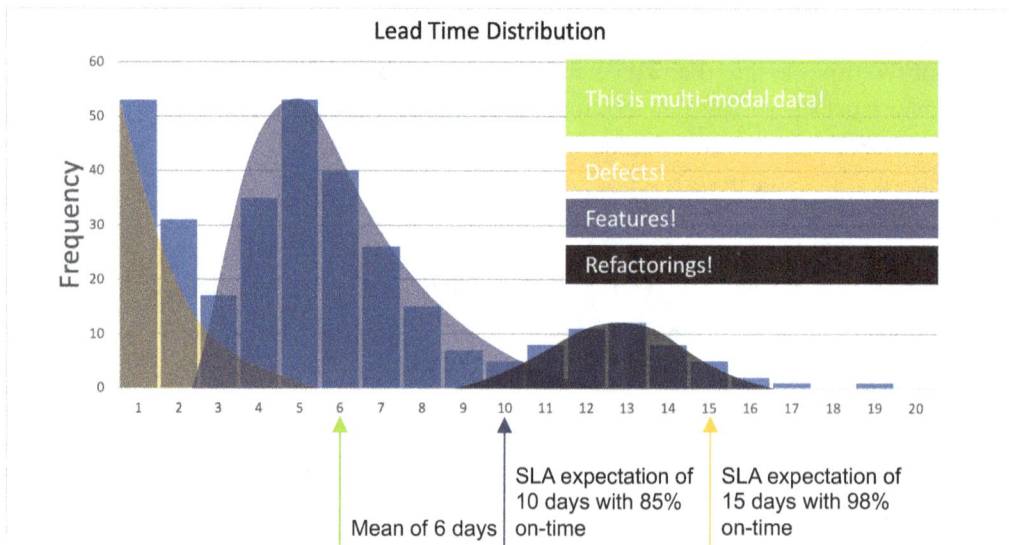

Figure A.4 Multi-modal data

If a curve has several peaks, it is said to be multi-modal. When this occurs in a kanban system's lead time distribution, it is usually an indication of multiple work item types. In the example in Figure A.4, there are three types: defect fixes, new features, and code and systems architecture refactoring. When this occurs, it is much better to filter the data set for only one work item type and obtain multiple lead time curves, each of which has only one peak and is therefore single modal.

Figure A.5 shows how the data in Figure A.4 look once filtered for each type of work.

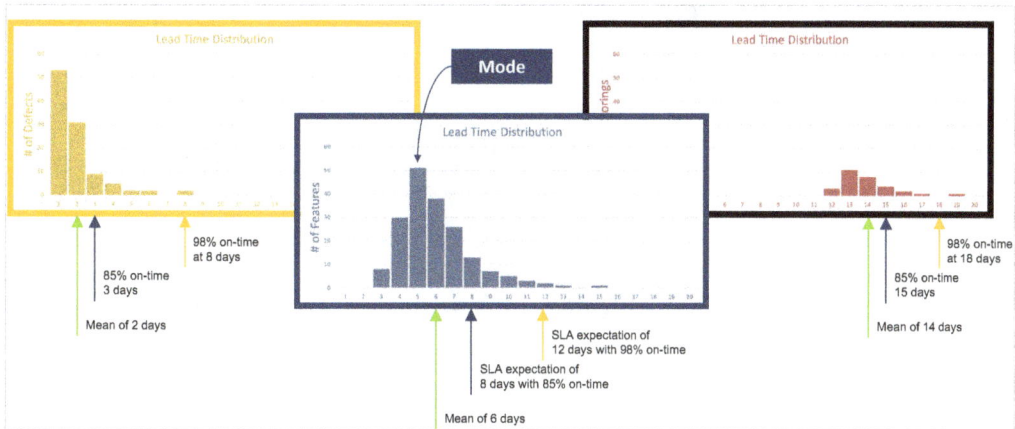

Figure A.5 Filtering the data by work item type

It is important to be able to report the current capability for a kanban system. A single number isn't an acceptable answer. Three pieces of data are required:

1. The mode to median: Most items are completed in approximately fifteen (mode) to twenty (median) days.

2. A high percentile, such as the 85th: Six out of seven items are completed within thirty (85th percentile) days.

3. Is it a thin- or fat-tailed distribution?

In order to use functions of averages such as Little's Law for forecasting or capacity allocation, it is necessary that the concept of an average is a meaningful number within a pragmatic quantity of data points. This is not the case for a fat-tailed distribution. Fat tails exclude us from using techniques such as capacity allocation and linear regression of delivery rate for forecasting. The mathematical explanation behind these observations follows.

Mathematical Properties of Lead Time Curves

As hinted at above in the advice for recognizing thin-tailed versus fat-tailed lead time distributions, lead time curves have been observed to have a set of mathematical properties that map from the rather Gaussian range of bell curve–like functions through to the Pareto range of power-law functions. The work of Troy Magennis revealed that the Weibull function provided a very useful means to map this range of curves using a single parameterized equation. This groundbreaking discovery proved immensely useful in helping us understand the mathematical properties of lead time and as a means to classify and categorize risk. It gave us a way to understand when equations of averages are appropriate, both in planning and forecasting and in product and risk management. This section explains the mathematical properties of Gaussian and Pareto curves, as well as the Weibull function, which integrates both into a single spectrum.

The Weibull Function

The Weibull distribution function, shown in Figure A.6, is named after Swedish mathematician Waloddi Weibull.[3] It turns out to be exceptionally useful, allowing us to map an entire universe of risk and probability distributions. The Weibull function enables us to model the entire set of domains that we see in project, program, portfolio, service delivery, and product management, namely, Gaussian, Super-exponential, and Pareto, by varying the shape parameter kappa (κ):

$$2.0 < \kappa <= 4.0 \text{ Gaussian range}$$
$$1.0 < \kappa <= 2.0 \text{ Super-exponential range}$$
$$\kappa = 1.0 \text{ Exponential function}$$
$$\kappa < 1.0 \text{ Pareto range}$$

$$f(x; \lambda, \kappa) = \begin{cases} \frac{\kappa}{\lambda}\left(\frac{x}{\lambda}\right)^{\kappa-1} e^{-(x/\lambda)^{\kappa}} & x \geq 0, \\ 0 & x < 0, \end{cases}$$

Figure A.6 Weibull distribution function

Further, we like to divide the Pareto range into near-exponential, or sub-exponential, and the more extreme power laws of the full Pareto range. For example, a sub-exponential distribution may have a mode of 1 and a tail stretching to 100 (or 1×10^2). In this range, using established project management and process improvement techniques, it's possible to affect the distribution, modifying it to become super-exponential. More extreme Pareto distributions may have tails with values such as 1×10^{13}. In such an extreme domain, it is not possible to hack the distribution function into the super-exponential range, and an entirely different set of risk-management approaches must be deployed. Figure A.7 shows four domains, breaking the Pareto range into the sub-exponential (hackable) range and the extreme power law range.

In an ideal world, where a worker is never pre-empted, never interrupted, never multi-tasks, and is never blocked, the length of time to complete pieces of work of a similar type is Gaussian distributed. However, the real world isn't like that. Workers are interrupted and work does get blocked and delayed. As there is no such thing as negative delay, all delays add to the time needed to complete work, and the consequence is that all lead time distributions are right skewed. In a mature, well-managed kanban system with smooth flow, the lead time distribution tends to be in the super-exponential range, with a kappa (κ) in the range 1.4 to 1.6. We often use $\kappa = 1.5$ as a good approximation or guess for the shape of a Maturity Level 3 or 4 kanban system. In the right-hand chart of Figure A.1, $\kappa = 1.4$.

3. https://en.wikipedia.org/wiki/Weibull_distribution

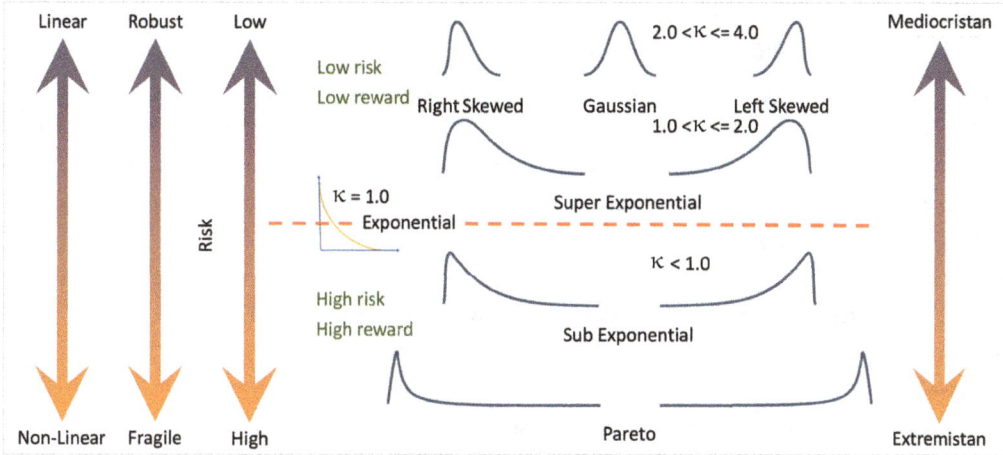

Figure A.7 Four domains of Pareto distributions

In less mature workflows, Maturity Level 2, the lead time tends to be sub-exponential, as flow is not well managed. In complex networks of interdependent services with lots of dependencies, work tends to get delayed often and the lead time tends to be sub-exponential. In the left-hand chart of Figure A.1, κ = 0.8.

To better understand the risks and what management practices are needed to mitigate them, we need to understand a little more about the mathematical properties of these ranges of probability distributions.

Gaussian Range

The key property of the Gaussian range is that both the head and tail of the function are asymptotic to the x axis (Figure A.8). Gaussian functions depart from the x axis and arrive back to the x axis. In a perfect bell curve, κ = 4.0. In skewed curves, kappa is in the range 2.0 < κ < 4.0; closer to 2.0 creates a greater skew.

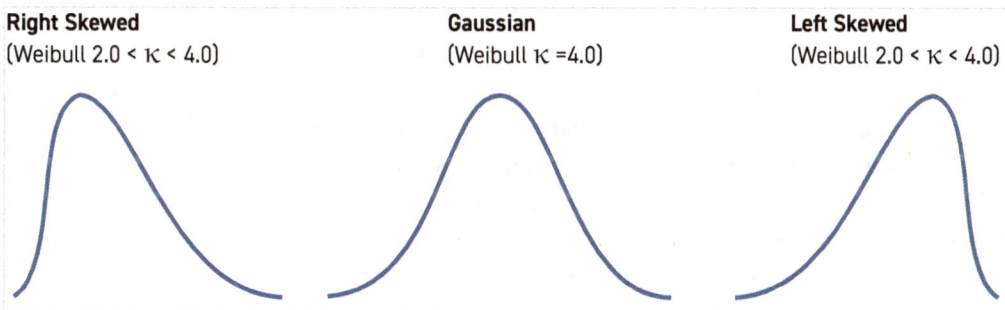

Figure A.8 Gaussian curves

The central limit theorem (also known as the law of large numbers) applies and is useful in this range. Regression to the mean within a 5 to 10 percent error is typically possible with fifteen to thirty data points, depending on the amount of skew. Gaussian-distributed data, on average, produces a linear line. Hence, linear regression is a reasonable technique. In risk management, Gaussian-distributed data sets and linear payoff functions, or risk functions, are synonymous. This is why Gaussian distributions are considered boring—Mediocristan.

The pro of Gaussian data is that it is easy to work with, easy to model, and simple to forecast with linear regression. The con, however, is counterintuitive. We need a large number of data points to have confidence in the range, or spread, of the function—its alpha—and to have confidence in the value of the mean.

Figure A.9 shows the spread of the Gaussian curve and its attribute alpha, α, describing its spread or range. The greater the alpha, the more data points are needed to regress within a reasonable variance of the mean. Models of real-world phenomena that exhibit Gaussian properties are sensitive to the value of alpha. Forecasts can be damagingly inaccurate if the wrong assumptions for mean and alpha are made. It can be necessary to have 1,000 to 2,000 data points in order to have strong confidence in the value of alpha.

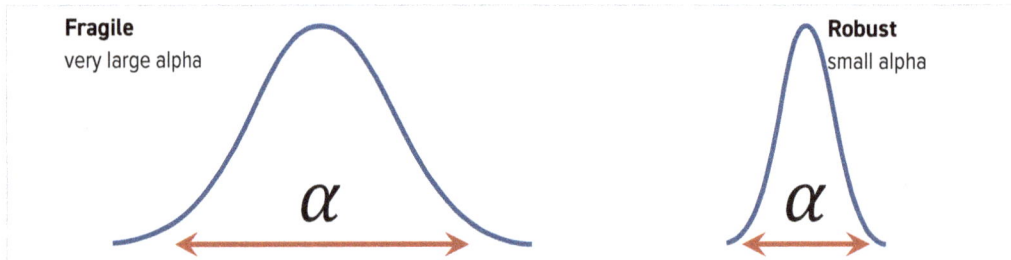

Fragile
very large alpha

α

Robust
small alpha

α

Figure A.9 Alphas of Gaussian curves

This creates an interesting dichotomy for Gaussian data; we need very few current data points to model and forecast, and the forecasting can be done with simple algebra. However, we need a much larger set of data points to have confidence in our model—our estimate of the shape of the curve. This makes working with Gaussian data problematic in real life. There is risk of model error—risk that we assume a narrow alpha when, in fact, we have a much broader one. When that happens, project plans are wildly inaccurate.

In real life, we know that kanban system delivery rates (also known as productivity rate, velocity, or throughput rate) are Gaussian distributed. We also know that the liquidity of a kanban system, the volatility in the rate of pull within the system, is Gaussian distributed. So, the rate of movement of work items is Gaussian at both a fine-grained level within the system and at the whole system level. This is a useful property when it comes to sizing input buffers, or reservation systems, or dependency buffers (parking lots).

Typical Gaussian functions observed in service delivery and project management look like the one in Figure A.10, right skewed and exhibiting a spread of two times above and below the mean.

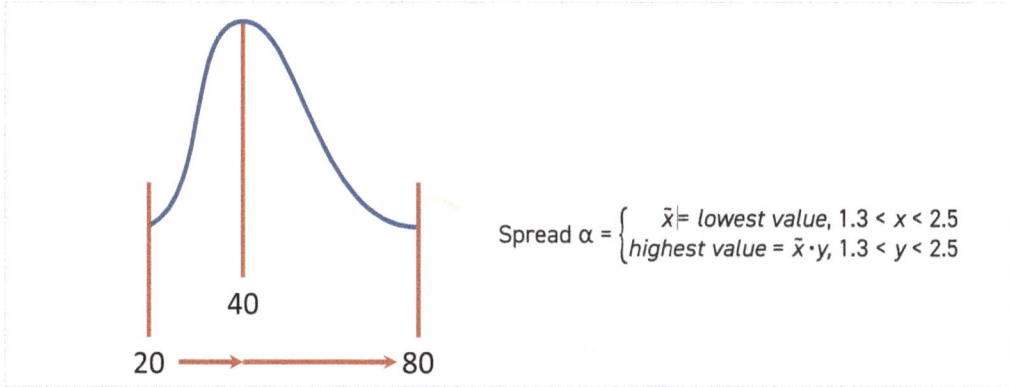

$$\text{Spread } \alpha = \begin{cases} \bar{x} = \text{lowest value, } 1.3 < x < 2.5 \\ \text{highest value} = \bar{x} \cdot y, \ 1.3 < y < 2.5 \end{cases}$$

Figure A.10 Right-skewed Gaussian curve

(Service) Delivery Rates typically exhibit an alpha in a range where the median = $x \times$ lowest value, $1.3 < x < 2.5$, and the highest value = median $\times y$, $1.3 < y < 2.5$. For alphas in this range, sixteen to thirty data points are sufficient for the central limit theorem to produce a relative error in regression to the mean of less than 10 percent. Or, linear regression of the mean value, twenty to twenty-five data points into the future are sufficient to produce a plan that is "fit-for-purpose." #noEstimates works when you project approximately twenty-five data points into the future. If you have weekly velocity data, then a forecast over six months (twenty-six data points) will be accurate using linear regression; there is no added benefit from a Monte Carlo simulation.

Super-Exponential Range

The key property of the super-exponential range is that both the head and tail of the function are asymptotic to different axes, the head to the y axis, the tail to the x axis. Super-exponential functions depart from the axis and arrive back to the x axis. They have kappa in the range $1.0 < \kappa <= 2.0$, as shown in Figure A.11. A kappa closer to 1.0 has a longer, fatter tail. Super-exponential functions aggregate to Gaussian with enough data points, and the concept of an average, or mean, is still meaningful.

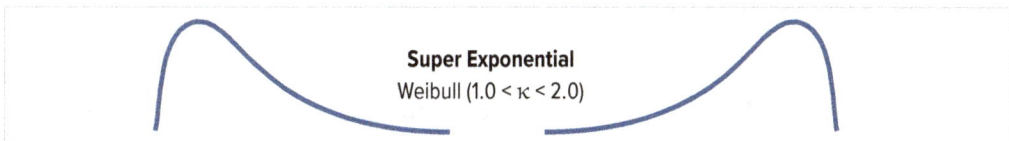

Super Exponential
Weibull ($1.0 < \kappa < 2.0$)

Figure A.11 Super-exponential range of Gaussian function

As shown in Figure A.11, the mode to tail ratio should be less than 16 and the median to tail ratio should be less than 6, assuming a location of 0. On the pro side, it is less sensitive to errors in alpha and hence, useful for modeling from relatively small data sets—eleven historical data points are often good enough, approximately thirty data points produce robust results with little model error (i.e., we have high confidence in the shape of the curve). The con is that linear regression of mean isn't useful until you have around eighty or more data points. Hence, it is riskier to use it with Little's Law—an equation of averages. For example, if we plan to use Little's Law to establish a capacity allocation, and on a monthly basis we need to meet our customer expectations about six months out of every seven, then we would want to know that the average throughput through that lane of the kanban board is at least eighty items per month. This would give us robust, predictable, trustworthy results. If we have lower throughput, we expose ourselves to some fragility—some risk that we will let customers down more often than is desirable.

Lead time distributions from well-managed kanban systems are usually super-exponential. Kanban system lead times tend to have $\kappa \sim= 1.5$. Troy Magennis has reported that Scrum lead times measured only within a sprint tend to have $\kappa \sim= 1.2$ (longer tailed, riskier than kanban). It is unclear what might cause this phenomenon at what is essentially a Maturity Level 1, team level, workflow. The most reasonable explanation is a lack of "focus." In Scrum, "focus" is a proxy for controlling WIP. Guidance is that a team shouldn't start too much all at the same time. A lack of focus increases multitasking and hence, fattens the tail of the lead time distribution. Note: this data reported by Troy Magennis did not reflect customer lead time, only local system lead time within a Scrum sprint tracking system.

In general, this super-exponential curve shape is associated with lead times for work with few external dependencies.

Figure A.12 shows the results of a simulation performed by Alexei Zheglov, mapping the median sample error for regression to the mean of a super-exponential distribution with $\kappa = 1.5$, using the popular and well-established central limit theorem and a new formula from Nicholas Nassim Taleb, known as the pre-asymptotic formula. The pre-asymptotic formula assumes that you have data points—at either end of the function—in the asymptotic range, close to the axes. In other words, it is a more conservative function, assuming that the set of data points you have so far is from the more extreme part of the range you might expect. The y axis shows the median error, and it is desirable to get this below 10 percent for psychological reasons, for example:

> "How long will my project take to complete?"
> "50 days."

Our experience of executive tolerance for error in such promises is 5 to 15 percent and hence, 10 percent is the midpoint, or median executive tolerance threshold. So, we want to know how many items need to be in the project in order for our forecast to lie in the forty-five to fifty-five days range, or within a 10 percent error of our estimate of fifty days.

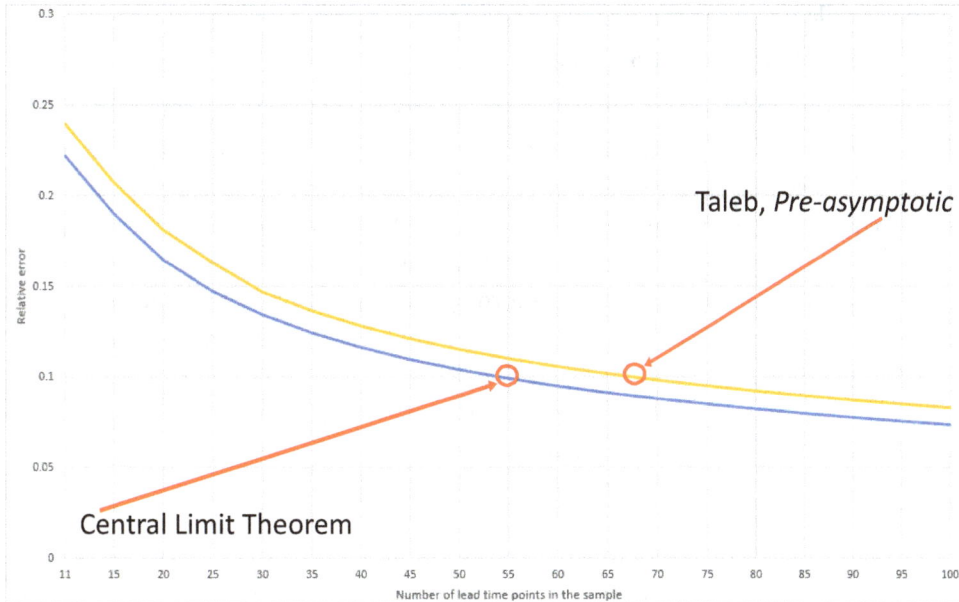

Figure A.12 Relative lead time averaging error

The result of the simulation in Figure A.12 suggests that fifty-five data points are needed with the central limit theorem. The more conservative pre-asymptotic formula produces a result of sixty-eight. Given that the kappa of the super-exponential function may be lower than $\kappa = 1.5$, our general guidance is that eighty to one hundred data points are needed for regression to the mean to be a useful technique when lead times are distributed in the super-exponential range. It is possible to trade down to twenty-nine data points for just a 5 percent relative error (i.e., +/−15 percent of the mean with half the data points).

Recommendation In the super-exponential range, you need seventy to one hundred data points for the concept of an average to be a meaningful value with an acceptable psychological tolerance for error.

Recommendation Little's Law is valid as a forecasting tool for kanban systems exhibiting a super-exponential lead time probability distribution, providing you expect at least 70 tickets to flow through your kanban board within the period of time that you are forecasting.

This is a highly important result! It means that Maturity Level 3 or deeper kanban systems exhibiting thin-tailed lead time distributions in the super-exponential range can be used to facilitate:

- Batch processing forecasts (e.g., project schedules) using linear regression of the delivery rate
- Buffer sizing, including input buffers and parking lots for dependencies

- Capacity allocation for demand shaping or risk hedging
- Sizing of booking/reservation systems
- Capacity allocation for classes of booking/reservation in a dynamic reservation system

Alexei Zheglov has described the super-exponential range as the sweet spot of project management forecasting, because only eleven to thirty data points are needed for extremely high model confidence, while fewer than one hundred data points are needed for high confidence in simple arithmetic forecasting techniques such as linear regression. He has also gone on to say that by far the most important thing any organization can do to improve its resilience—its customer satisfaction, its fitness-for-purpose, its robustness—and manage risk is first to instrument for lead time and then to focus on trimming the tail of the lead time distribution. This is the job of the Flow Manager role at Maturity Level 2, and it is vital to achieving Maturity Level 3 and deeper.

Approximately Exponential

The exponential function can be modeled with the Weibull equation using kappa, κ = 1.0. It represents the inflection point between the world of Mediocristan—the super-exponential and Gaussian probability distribution range—and the world of Extremistan, the world of Pareto distribution and power laws. Even close to the exponential function, simple forecasting is possible if we are willing to trade a little relative forecasting error . . .

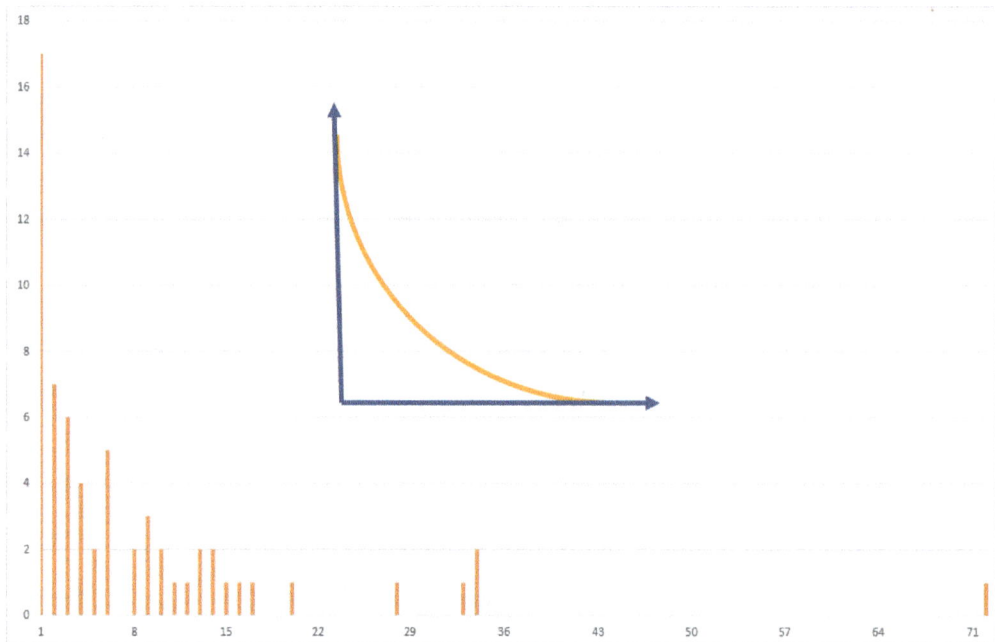

Figure A.13 Real-world data compared with classic exponential function

Figure A.13 shows a real-world data set, with the canonical exponential function inlaid in the picture. Excluding the outlier data point of seventy-two days, which originated from the bootstrap of the project and therefore does not come from a period of equilibrium and stable volatility, the remaining data points exhibit a distribution very close to, or an approximation of, the exponential function.

Figure A.14 shows the cumulative flow diagram for the same project.[4] For 150 data points, observation gives us 150 items in 154 days, or .97 per day, and Little's Law gives us 0.83 per day for an average WIP of 5.2 and observed lead time mean of 6.3, or a 15 percent relative forecasting error.

Recommendation Simple arithmetic forecasting remains possible at or close to the exponential distribution for lead time when the number of data points—the number of work items flowing through the kanban system—exceeds 150 in the relevant time period. The corollary of which is to say that for exponential-, super-exponential-, and Gaussian-distributed data sets, algorithmic, iterative techniques such as Monte Carlo are not required.

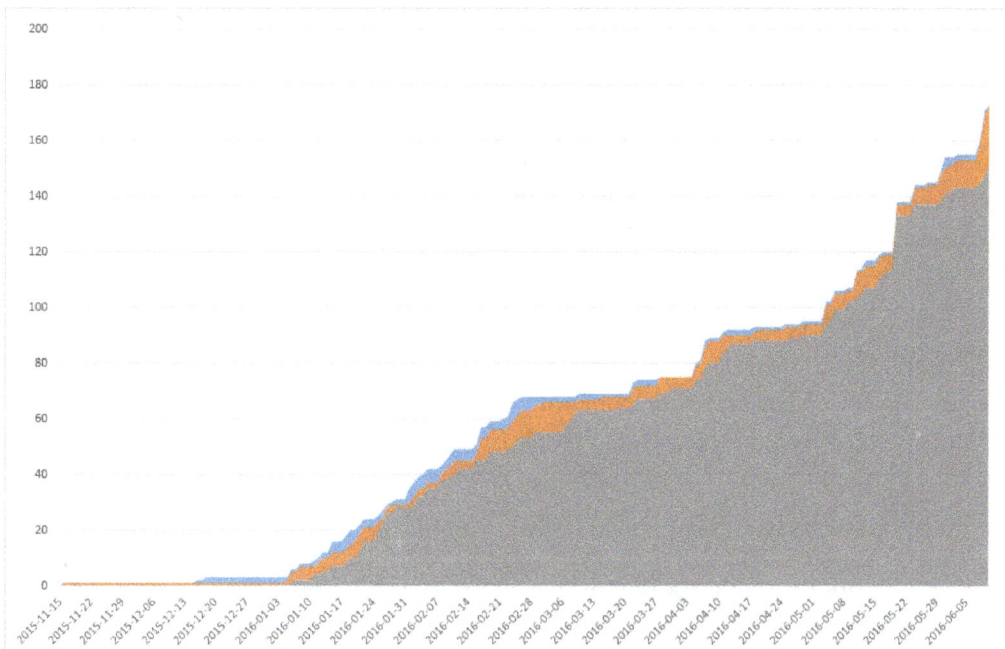

Figure A.14 Cumulative flow diagram

---

4. Both A.13 and A.14 courtesy of Andreas Bartel, flow.hamburg

Sub-Exponential Range

The sub-exponential range, shown in Figure A.15, is part of the Pareto range of power laws. We are choosing to break it out separately as we believe that the range of power laws, relatively close to the exponential function, are "hackable." In other words, there are pragmatic actions we can take to trim the tail and hack the distribution into the desirable super-exponential range.

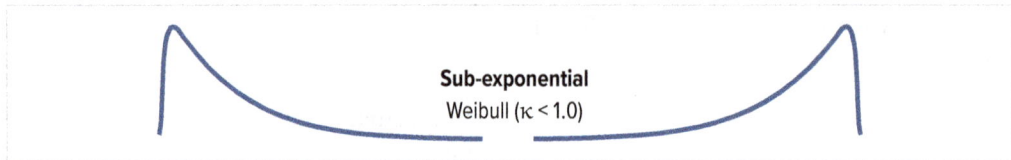

Sub-exponential
Weibull ($\kappa < 1.0$)

Figure A.15 Sub-exponential range

Sub-exponential functions are modeled as Weibull, with kappa in the range of $0.7 < \kappa < 1.0$. We consider these semi-fat tailed because the tail can be hacked through our actions. The median to tail ratio is typically in the range of six to one hundred times. If the median is five, we expect a tail in the range of thirty to five hundred, and not measured in thousands or tens of thousands or even higher. Lead time distributions for IT operations work are often sub-exponential. In general, this shape is associated with lead times for work with many external dependencies.

The pro for this function is that it is even less sensitive to errors in alpha and hence, is useful for modeling from relatively small data sets—eleven historical data points should produce robust results for estimating the shape of the curve. However, the con is that linear regression of the mean isn't useful at all. At around eighty data points it fails to get within 20 percent of the mean. Alexei Zheglov's simulation, plotted in Figure A.16, demonstrates that at least 2,000 data points are required to reduce the error to within psychological tolerances. Because of this, fat-tailed data sets should not be used with Little's Law—an equation of averages.

Recommendation In the sub-exponential range, arithmetic forecasting techniques, such as linear regression and the use of Little's Law, are not valid for buffer sizing, capacity allocation, reservation system sizing, and dependency parking lots. Although sub-exponential range data sets are suitable for iterative simulation techniques such as Monte Carlo, we do not believe that the range of outcomes from the simulations represent useful, pragmatic, actionable guidance. For example, simulating a range of WIP limits to use for a capacity allocation on a kanban board is not useful. In a context where a single number is needed, we need to be able to depend on the arithmetic mean. Therefore, the guidance is simple: Focus on trimming the tail of the lead time distribution using techniques such as blocker clustering, root cause analysis, risk mitigation and reduction, and reducing multitasking and pre-empting of work that causes delay. Reduce and limit WIP to discourage multitasking and pre-empting, and focus on quality and first-time right with the intent of avoiding rework and kanban tickets that cycle backward on the board, which result in longer lead times. In other words, use specific practices from Maturity Level 2 and Maturity Level 3 to control and eliminate the tail and hack the distribution into the super-exponential range.

Figure A.16 shows a chart, similar to the one in Figure A.12, displaying the mean relative error for regression to the mean, this time for a sub-exponential Weibull distribution with kappa, $\kappa = 0.8$. At one hundred data points, the median error is only just under 20 percent. In other words, if you forecast that a project with one hundred items to deliver takes fifty days, you have only a one in two chance that it will take between forty and sixty days, and you have a one in four chance that it will take more than sixty days. These are not good odds, psychologically speaking. It is unlikely that you can operate a trustworthy service delivery with such a forecast. This simulation showed extreme asymptotic behavior to the x axis and failed to fall below a 10 percent median error until it had 2,000 data points. Raphael Douady[5] advised Alexei Zheglov that 10,000 data points may be needed to make safe estimates and for the concept of a mean to be a meaningful idea.

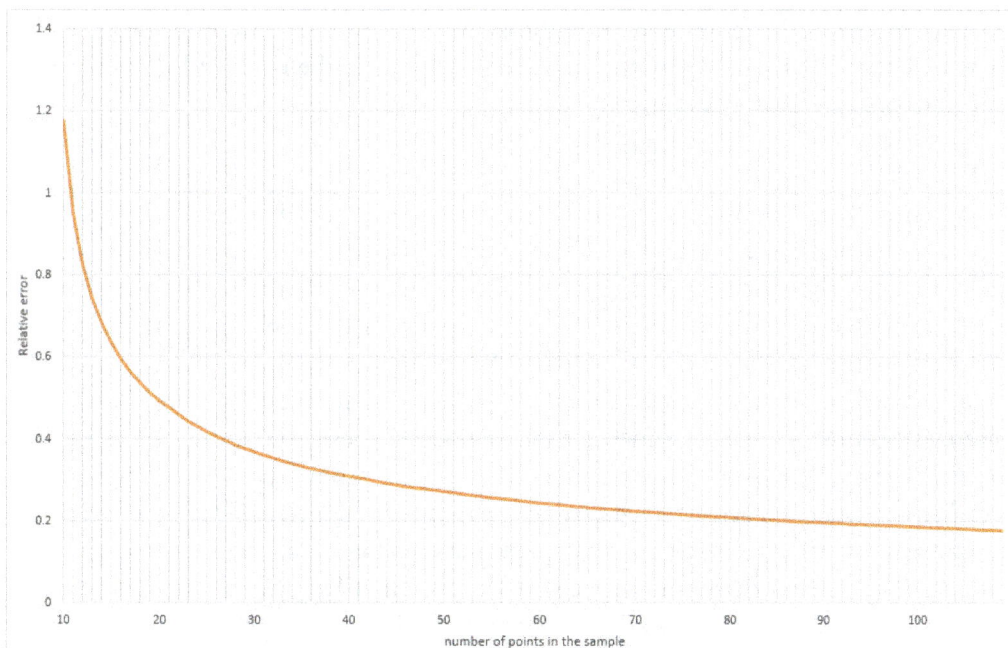

Figure A.16 Relative lead time averaging error estimator (Weibull [.0.5 < κ < 1.0])

5. https://en.wikipedia.org/wiki/Raphael_Douady

The simple advice is that if you find yourself in the sub-exponential domain, averages are meaningless. Any technique or equation that requires an average is invalid. The implications of this are quite profound, as it implies that pretty much all published and well-established guidance on project management, portfolio management, and product management does not work in Extremistan (the sub-exponential domain). For example, the standard project management and product management guidance on prioritization by return on investment using the equation

$$ROI = \frac{Value}{Cost}$$

simply cannot be used if there is any possibility that either the value or the cost is sub-exponentially (Pareto) distributed. In domains such as publishing, music publishing, computer games, and almost any market with a network effect, such as social media platforms, operating systems, and business productivity tools, such approaches to prioritization are invalid. An equation such as ROI implicitly assumes average value and average cost. There is an implicit assumption of Gaussian distribution for a range of value, and a range of cost, and that these ranges regress to the mean within a reasonable number of data points for the given market or business domain.

This text is not the forum to explore the answers to this problem. If this interests you, we encourage you to follow new and future publications in the fields of Fit-for-Purpose and Enterprise Services Planning.

Pareto Range

Pareto class distributions are often referred to as power laws. These are very fat tailed and represent high risk. Hence the nickname Extremistan. In this range, tails exceed one hundred times the mode, ranging out to values such as 10.[13] Although these distributions can be hacked through action, it is almost impossible to hack them into the super-exponential range. For example, the Pareto-distributed movie industry uses hacks such as the star system, franchises, sequels, and cross-licensing from books, comics, and other entertainment media such as games, theater, opera, and music to improve the chances of success. However, of all the movie franchises, only James Bond has successfully crossed into Mediocristan and achieved a sustainable concept of an "average" James Bond movie in terms of box office receipts, one that spans punctuation points, with changes of actor and script writers. For those who have counter-argued that *Star Wars* is a similar franchise that has crossed into Mediocristan, we counter with *Solo*, officially the first movie in the *Star Wars* franchise to flop.[6] As general advice, you can't hack the Pareto range into the Gaussian range, and hence, an entirely different approach to risk management is needed.

---

6. https://www.vanityfair.com/hollywood/2018/06/solo-star-wars-box-office-flop

Fortunately, Pareto distributions rarely, if ever, appear in project management or service delivery problems. Pareto distributions do exist in product management problems such as payoff uncertainty; for example, how much will our next book make? Is it the next *Harry Potter*? Or will it sell fewer than 300[7] copies? Pareto distribution problems are also known as black swan[8] problems. Project management and service delivery domains don't involve black swans, but product and portfolio management can.

---

7. Approximately 300 copies is the mean total sales of any specific book title on Amazon.com.
8. According to Taleb, a black swan event is an event that almost never happens, but when it does, it has major consequences. https://en.wikipedia.org/wiki/Black_swan_theory

Index

Acknowledgments

The famous quote by Bernard of Chartres—that we are just *dwarves standing on the shoulders of giants*—couldn't be more true for this book. This book wouldn't be possible without those studying the Upstream Kanban concepts (more or less explicitly) before me: Patrick Steyaert with *Essential Upstream Kanban* and the SCIA case study, Chris Matts and Olav Maassen with *Commitment*, and, obviously, Donald G. Reinertsen with the classic *The Principles of Product Development Flow*—the book that influenced my thinking long before I ever heard about the Kanban Method.

I'm very grateful for the non-obvious path in my life that brought me here through my career as an accountant and my two great managers: Renata Rybak and Paulina Handke, who were always supportive and motivated me to pursue new goals and who helped me to develop even the craziest ideas.

The person who showed me the Kanban Method—Katarzyna Banaś. I wouldn't be here without you!

The previous team of awesome people who were never afraid of trusting me when I was coming to them with new improvements (some of them being rather experiments than improvements): Martyna Bednarska, Piotr Gąsiorowski, Bartłomiej Bujak, Monika Kozak, Magda Michałek, Krzysztof Pluszyński, and Łukasz Klepek.

My current teams at the David J Anderson School of Management, Mauvius Group Europe, and Kanban University: Lauren Durdan, Jan Radzikowski, Joanne Madeja, Nastya Kondratova, Anya Kondratova, Diana Deriazhna, John Kennedy, Monica Iturriaga, and Todd Little. Thank you for sticking together through storms and sunny days—so similar to the Basque weather!

Special shout out to the one and only Stephanie Dziad, who is the best friend of each Kanban trainer and whom I have the honor to call my friend. My Kanban life wouldn't be the same without you!

A separate big thank you goes to the students who took the *Kanban for Design and Innovation* class and shared their experiences and feedback with us! Each of your insights makes the material more pragmatic, actionable, and evidence-based.

And, last but not least, the best boss and mentor I've ever had, David J Anderson. I'm immensely proud and grateful to have had a chance to work with you!

www.ingramcontent.com/pod-product-compliance
Lightning Source LLC
Chambersburg PA
CBHW080553220326
41599CB00032B/6467